Yale Mathematical Monographs 4

James K. Whittemore Lectures in Mathematics
given at Yale University

# GROUP THEORY AND THREE-DIMENSIONAL MANIFOLDS

by John Stallings

New Haven and London, Yale University Press, 1971

Library of Congress catalog card number: 70-151590.

International standard book number: 0-300-01397-3 (paper).

Set in IBM Bold Face One type.
Printed in the United States of America by
The Murray Printing Co., Forge Village, Mass.

Distributed in Great Britain, Europe, and Africa by Yale University Press, Ltd., London; in Canada by McGill-Queen's University Press, Montreal; in Mexico by Centro Interamericano de Libros Académicos, Mexico City; in Central and South America by Kaiman & Polon, Inc., New York City; in Australasia by Australia and New Zealand Book Co., Pty., Ltd., Artarmon, New South Wales; in India by UBS Publishers' Distributors Pvt., Ltd., Delhi; in Japan by John Weatherhill, Inc., Tokyo.

CONTENTS

# 1 BACKGROUND AND SIGNIFICANCE OF THREE-DIMENSIONAL MANIFOLDS

## 1.A Introduction

The study of three-dimensional manifolds has often interacted with a certain stream of group theory, which is concerned with free groups, free products, finite presentations of groups, and similar combinatorial matters.

Thus, Kneser's fundamental paper [4] had latent implications toward Grushko's Theorem [8]; one of the sections of that paper dealt with the theorem that, if a manifold's fundamental group is a free product, then the manifold exhibits this geometrically, being divided into two regions by a sphere with appropriate properties. Kneser's proof is fraught with geometric hazards, but one of the steps, consisting of modifying the 1-skeleton of the manifold and dividing it up, contains obscurely something like Grushko's Theorem: that a set of generators of a free product can be modified in a certain simple way so as to be the union of sets of generators of the factors.

Similarly, in the sequence of theorems by Papakyriakopoulos, the Loop Theorem [14], Dehn's Lemma, and the Sphere Theorem [15], there are implicit facts about group theory, which form the main subject of these chapters.

Philosophically speaking, the depth and beauty of 3-manifold theory is, it seems to me, mainly due to the fact that its theorems have offshoots that eventually blossom in a different subject, namely group theory. Thus I

A James K. Whittemore Lecture, October 1969.

tend to believe that new results in the theory, such as Waldhausen's [26], may eventually have relatives in group theory; the solution of the Poincaré Conjecture [1], if it ever occurs, will have group-theoretic consequences of a wider nature.

To see the connection between the results of Papakyriakopoulos and group theory, let us consider the Sphere Theorem [15]. This says:

1.A.1 If $M$ is an orientable 3-manifold, with nonzero $\pi_2(M)$, then there is a 2-sphere $S$ embedded in $M$ which is not contractible in $M$.

A nonorientable version also exists [22]:

1.A.2 If $M$ is a 3-manifold, orientable or not, with nonzero $\pi_2(M)$, then there is a 2-sided 2-manifold $\Sigma$ in $M$, where $\Sigma$ is either a 2-sphere or a projective plane carrying a nonzero element of $\pi_2(M)$.

The existence of a noncontractible 2-sphere in $M$ implies, if $M$ is compact and boundariless, that $\pi_1(M)$ is either infinite cyclic or a nontrivial free product. The group-theoretic meaning of the existence of a 2-sided projective plane is harder to state since there is no standard terminology for the situation of groups found here; we shall momentarily have to be content to say that $\pi_1(M)$ is rather like a free product with amalgamated subgroup of order two.

Thus the sphere theorem implies that if a group can be represented as $\pi_1(M)$, for a certain sort of space $M$, and $M$ satisfies $\pi_2(M) \neq 0$, then we can decompose the group into a free product, perhaps with a certain finite amalgamated subgroup.

A similar situation occurs in Dehn's Lemma and the Loop Theorem. These combine to the following theorem:

1.A.3 If $M$ is a 3-manifold with boundary $B$ and the kernel of the injection homomorphism $\pi_1(B) \to \pi_1(M)$ is nontrivial, then there is a 2-cell $\Delta$ in $M$ whose boundary lies in $B$ and is not contractible there.

The conclusion of this theorem implies, as does the Sphere Theorem, that, under suitable hypotheses on $B$, $\pi_1(M)$ is infinite cyclic or a nontrivial free product. The hypothesis can be reformulated, taking notice of the exact sequence of homotopy groups of the pair (M, B), leading to the following group-theoretic conclusion:

1.A.4 If G is $\pi_1(M)$ where M is an orientable (or nonorientable) 3-manifold with boundary B, no component of which is a 2-sphere or projective plane, and if $\pi_2(M, B) \neq 0$, then G is either infinite cyclic or a free product (perhaps in some sense with a $Z_2$-amalgamation).

The proofs of Dehn's Lemma and the Loop Theorem are an order of magnitude easier than is the proof of the Sphere Theorem. It may therefore be of interest to note that the Sphere Theorem is a consequence, by a fairly simple argument, of our main group-theoretic result and the easier theorems of Papakyriokopoulos. We give an indication of how this argument goes in 5.C.

The fact that the hypothesis of the above result involves $\pi_2(M, B)$ instead of something connected with G is remedied thus [12]. We use Poincaré duality in the universal covering space of M and the assumption that B is aspherical to see that $\pi_2(M, B) \neq 0$ if and only if $H_f^1(\tilde{M}) \neq 0$, where $\tilde{M}$ is the universal covering space of M, and $H_f^1$ denotes the cohomology group computed with finite cochains. This latter group is closely connected with $\pi_1(M)$ by homological conditions.

We now say something about ends. If X is a connected, locally connected, locally compact Hausdorff space, then the set of ends of X is the inverse limit of the directed set whose typical element is the set of components of X − C, where C is any compact subset of X. Thus the ends of the real line consist of two points, of the plane one point, of any locally finite, connected tree some closed subset of the Cantor set.

Ends are related to cohomology by various statements, of which the following is one:

1.A.5 If K is a connected, locally finite complex, then K has more than one end if and only if the kernel of $H_f^1(K) \to H^1(K)$ is nontrivial.

In the case at hand, the condition $\pi_2(M, B) \neq 0$ is equivalent to the condition that $\widetilde{M}$ have more than one end. The ends of $\widetilde{M}$ are determinable in an algebraic way from $\pi_1(M)$ and lead to the notion of the ends of an abstract, finitely generated group. Therefore, the natural conjecture is the following:

1.A.6 A finitely generated group G with more than one end is either infinite cyclic or a nontrivial free product or something analogous with finite amalgamated subgroup.

This we shall state precisely and prove and thereby completely characterize, in a sense, finitely generated groups with more than one end.

## 1.B Precise Statements of Quoted Theorems

By a 3-manifold, we mean a Hausdorff space such that every point has a neighborhood homeomorphic either to 3-dimensional Euclidean space or to a closed half-space thereof. However, we always deal with the polyhedral context, involving the underlying space of a simplicial complex K whose vertices have as links either 2-spheres or 2-cells; our techniques are always polyhedral (such as general position and cut-and-paste). These two pictures are more or less equivalent for metrizable 3-manifolds, thanks to papers of Moise [13] and Bing [19].

The precise statement of the Sphere Theorem of Papakyriakopoulos [15] as refined by Whitehead [17] and Epstein [22] is:

1.B.1 Sphere Theorem: Let M be a 3-manifold and let A be a $\pi_1(M)$-submodule of $\pi_2(M)$ such that $\pi_2(M) - A \neq \emptyset$. Then there is $X \subset M$, such that X is homeomorphic to the 2-sphere or to the real projective plane and such that X has a neighborhood in M homeomorphic to $X \times (-1, +1)$ and such that the fundamental element of $\pi_2(X)$ represents an element of $\pi_2(M) - A$.

The precise statement of the Loop Theorem and Dehn's Lemma, as proved by Papakyriakopoulos [14, 15] and refined by Shapiro and Whitehead [18] and Stallings [20] is:

# SUBJECT INDEX

1.B.2 Loop Theorem—Dehn's Lemma: Let M be a 3-manifold with boundary component B; let N be a normal subgroup of $\pi_1(B)$ such that

$$(\pi_1(B) - N) \cap (\text{kernel } \pi_1(B) \to \pi_1(M)) \neq \emptyset.$$

Then there is a 2-cell $D \subset M$ whose boundary is contained in B and there represents an element of $\pi_1(B) - N$.

A result important for our discussion is Kneser's Conjecture, a theorem proved by Kneser [4], using the unfortunately inaccurate techniques of Dehn [2], involving the notion of free product of groups. It is, as refined by Stallings [23] thus:

1.B.3 Kneser's Conjecture: Let M be a compact, connected 3-manifold with empty boundary; let $\phi$: $\pi_1(M) \to A * B$ be a homomorphism of $\pi_1(M)$ onto a free product such that, whenever T is a 2-sided 2-manifold in M and $\pi_1(T) \to \pi_1(M)$ is injective and $\phi(\pi_1(T)) = \{1\}$, then T is a 2-sphere. (This condition is always met if $\phi$ is an isomorphism or if $\pi_1(M)$ does not contain any nontrivial groups $\pi_1(T)$ for T a closed 2-manifold.) Then it is possible to write $M = M_A \cup M_B$, where $M_A \cap M_B$ is a 2-sphere and $\phi(\pi_1(M_A)) = A$, $\phi(\pi_1(M_B)) = B$.

It was at a stage in the proof of this result, for $\phi$ an isomorphism, that Kneser foresaw Grushko's Theorem. In fact, the result as stated above implies Grushko's theorem if we take M to be a 3-sphere with handles so that $\pi_1(M)$ is a free group.

A subset A of a topological space X is said to be bicollared if there is an open subset of X homeomorphic to $A \times (-1, +1)$, containing A as $A \times 0$. A result related to Kneser's Conjecture, and from which the sphere theorem can be derived eventually, states:

1.B.4 Let M be a compact 3-manifold, $\phi : M \to X$ a continuous function, where X contains a bicollared subset A. Assume $\phi_* : \pi_1(M) \to \pi_1(X)$ is injective and that $\pi_2(X - A \times (-\frac{1}{2}, +\frac{1}{2}), A \times \{-\frac{1}{2}, +\frac{1}{2}\}) = 0$ for all possible choices of base point. Then $\phi$ is homotopic to $f : M \to X$ such that $f^{-1}(A)$ is a bicollared 2-manifold contained in M for every component $T_i$ of which the homomorphism $f_* : \pi_1(T_i) \to \pi_1(A)$ is injective.

The Theorem of Waldhausen referred to is a bit too complicated to state here; given in reference [26], it has the consequence of classifying certain 3-manifolds by their fundamental groups alone (by means of a certain hypothesis getting around the all-important Poincaré Conjecture [1]).

We now say some words about group theory, reserving the details for chapter 3.

The free product $A * B$ of two groups A and B is the "coproduct" in the category of groups and homomorphisms, of A and B. It is classically described in terms of reduced words in $A \cup B$. With this description, free products were first considered by Schreier and Artin around 1925 and immediately generalized by Schreier [3] to the notion of free product with amalgamation. Such amalgamated free products arise topologically as the fundamental groups of the unions of pairs of spaces, according to the theorem of Seifert [5] and van Kampen [7].

A theorem on free products which is of special interest to us is Grushko's Theorem [8] (see also Neumann [9]), which in the form for infinitely generated groups is due to Wagner [16] (there is a topological proof by Stallings [24]). It states:

1.B.5 Grushko-Wagner Theorem: If F is a free group (i.e., a free product of infinite-cyclic groups) and $\phi : F \to *_\alpha A_\alpha$ a homomorphism onto a free product, then F can be written as a free product $F = *_\alpha F_\alpha$ such that $\phi(F_\alpha) = A_\alpha$.

Our work on ends of groups has led us to a generalization of the notion of free product with amalgamation which we discuss in some detail in chapter 3.

We first consider "pregroups." A pregroup is an algebraic system in which multiplication is not always defined but which is otherwise very like a group; there is a unit element, an inverse to each element, and associativity when possible; there is also a peculiar property that, whenever wx, xy, and yz are defined, then at least one of the triple products wxy or xyz is defined.

Pregroups have universal groups in which the word problem is solvable in a manner similar to that existing in free products with amalgamation. The universal group U(P) is defined as being in a certain sense the largest group that could be generated by the pregroup P. There is an equivalence relation on the reduced words in P which is generated by the example (x,y) is equivalent to $(xa, a^{-1}y)$ when the products involved are defined. The theorem is that every element of U(P) is represented by a unique equivalence class of reduced words.

A particular instance of the universal group of a pregroup is the free product with amalgamation.

We were led to consider pregroups by examining van der Waerden's proof of the associative law for free products defined in terms of reduced words [11]. Pregroups contain what seem to be the minimal assumptions necessary for this argument to work.

Pregroups are of use in classifying another sort of combinatorial group-theoretic situation, that of bipolar structures. Originally we defined bipolar structures for the case of torsion-free groups [27], but it seems wise to rewrite the concept so as to take care of periodic elements.

A bipolar structure on a group G occurs when G can be expressed as a free product of two groups A and B with finite amalgamated subgroup F. In this case, the elements of $G - F$ fall into four classes, depending on whether their equivalence class of reduced words in $A \cup B$ begins or ends in A or B. We can axiomatize the situation roughly as follows: We have a finite subgroup F and four subsets, denoted EE, EE*, E*E, E*E* which make a partition of G such that if X and Y stand for E or E* and we have the convention that $E^{**} = E$, then, if $g \in XY$ and $a \in F$, we have $ga \in XY$; and if $g \in XY$ and $h \in Y^*Z$, then $gh \in XZ$; and finally, given any $g \in G$, there is a bound to the lengths of expressions $g = g_1 g_2 \cdots g_n$, where $g_i \in X_i X_{i+1}^*$. For the general case, we have to allow also the possibility of another set of the partition, S, where $F \cup S$ is a subgroup in which F has index one or two, and such that for $g \in XY$ and $a \in S$ we have $ga \in XY^*$.

The structure of a group with a bipolar structure can be analyzed by noting that it is the universal group of its pregroup consisting of $F \cup S \cup$ {the indecomposable elements}. Here, an indecomposable element is an element of $XY$ which cannot be expressed as the product of an element of $XZ$ and an element of $Z^*Y$. We can then see that the pregroup involved is one of three types, which are themselves easily understood as giving rise to a free product with amalgamation on a finite subgroup or else to the similar situation in which the group G is generated by a subgroup A and one additional element x and the relations derived from some embedding of a finite subgroup F of A into A again; these relations are of the form $xfx^{-1} = \phi(f)$.

We call the bipolar structure nontrivial if there exists an element of $EE^*$. Our main group-theoretic result can now be stated:

1.B.6 Every finitely generated group with more than one end has a nontrivial bipolar structure and so can be described as a nontrivial free product with finite amalgamated subgroup or as the other type of group. Conversely, any group with nontrivial bipolar structure has two or infinitely many ends.

We now say some words about graphs.

If $\Gamma$ is a locally finite graph (i.e., a 1-dimensional complex), we can look at certain cohomology groups with the simplest coefficient group $Z_2$. The ordinary (infinite) cochains contain as subcochain complex the finite cochains; the quotient complex $C_e^*(\Gamma)$ is where "end" phenomena are seen. In particular, $H_e^0(\Gamma)$ is the "group of ends," whose rank, as $Z_2$-module, is the "number of ends" of $\Gamma$. It can be seen that $H_e^0(\Gamma)$ in fact inherits a Boolean algebra structure from that of $C^0(\Gamma)$, which is that of the algebra of all subsets of vertices of $\Gamma$. The "space of ends" is then classically the maximal ideal space of the Boolean algebra $H_e^0(\Gamma)$.

If G is a finitely generated group, generated by $T = \{t_1, \ldots, t_n\}$, then we define the graph $\Gamma$ of this situation to have vertices G such that g and $t_i g$ are connected by an edge $(t_i, g)$. If we define A as a G-module to consist of all subsets of G modulo all finite subsets (addition being symmetric difference), then we have the fact that $H_e^0(\Gamma) \approx H^0(G;A)$, this latter being interpreted as in the theory of cohomology of groups. Thus the ends of $\Gamma$ are independent of T and will be called the ends of G.

If $\Gamma$ is a connected, locally finite graph with more than one end, then there exist 0-cochains Q with finite coboundary $\delta Q$, such that neither Q nor the complement $Q^*$ is finite. Among such Q there are those whose coboundary has the smallest number of elements; these will be called narrow 0-cochains. The narrow cochains satisfy some nice lattice-theoretic properties; in particular there is, given any vertex v, a smallest narrow 0-cochain containing v. For such a smallest narrow 0-cochain, Q, the following fact holds:

1.B.7 If X is any narrow cochain, then at least one of the 0-cochains $Q \cap X$, $Q \cap X^*$, $Q^* \cap X$, $Q^* \cap X^*$ is finite.

This is a crucial graph-theoretic result, which has implications in group theory thus. We suppose that $\Gamma$ is the graph of a group G with respect to some finite generating set; G acts on the right on $\Gamma$, and so Qg is narrow whenever Q is; we apply the above result when $X = Qg$. We find then that, given such a Q, there are six possibilities for which ones of the sets $Qg \cap Q$, $Qg \cap Q^*$, $Q^*g \cap Q$, $Q^*g \cap Q^*$ are finite. When the second and third of these sets are finite, we say $g \in F$. When the first and fourth are finite, we say $g \in S$. When only the first set is finite, we say $g \in EE$. When only the second set is finite we say $g \in EE^*$. If the third is the only finite set, we say $g \in E^*E$. If the fourth is the only finite set, we say $g \in E^*E^*$.

It turns out that, if G has more than two ends, this partition of G is a nontrivial bipolar structure. If G has exactly two ends, then the structure of G has been discussed by various people, and in particular it is clear from their work that such a G has a nontrivial bipolar structure.

By means of these results, we prove a conjecture of Eilenberg and Ganea [29].

1.B.8 Any finitely generated group G having cohomological dimension 1 is a free group.

This has lately been improved by Swan [28], who shows how to eliminate the hypothesis "finitely generated."

We also derive a result conjectured by Serre [30], which has in turn been proved for infinitely generated groups by Swan:

1.B.9 If a finitely generated group G has no nontrivial element of finite order and contains a free subgroup of finite index, then G is a free group itself.

This is the most significant result of our work since it involves only simple group-theoretic concepts and yet entails delicate combinatorial facts. The structure of a group G having a free subgroup of finite index is not yet in any completed formulation if we allow the group to have torsion elements.

# 2 THREE-DIMENSIONAL MANIFOLDS

## 2.A The Loop Theorem and Dehn's Lemma

Here we summarize the paper listed in the references as [20].

2.A.1 If $V$ is a compact 3-manifold, and $H^1(V;Z_2) = 0$ (which is equivalent to: $V$ has no connected 2-sheeted covering space), then every component of $\partial V$ is a 2-sphere.

Proof: From Poincaré duality, $H_2(V, \partial V; Z_2) \approx H^1(V;Z_2) = 0$. From the universal coefficient theorem $H_1(V;Z_2)$ is dual to $H^1(V;Z_2)$ and hence $0$. From the exact homology sequence, $H_1(\partial V; Z_2) = 0$, from which the lemma follows.

2.A.2 Let $V$ be a compact 3-manifold, $B \subset \partial V$ a 2-manifold, $N$ a proper normal subgroup of $\pi_1(B)$; suppose $V$ has no connected 2-sheeted covering space. Then there is a 2-cell $\Delta \subset V$, with $\partial\Delta \subset B$ not representing an element of $N$.

Proof: Since $\pi_1(B)$ is generated by nonsingular loops on $B$, at least one of them does not represent an element of $N$; such a loop will, by 2.A.1, bound a 2-cell $\Delta \subset \partial V$.

2.A.3 Let $f : \Delta \to K$ be a simplicial map, where $\Delta$ is a 1-connected finite complex, and $K = f(\Delta)$. Suppose we have a sequence of connected covering spaces:

$$P_i : L_{i+1} \to K_i, \quad \text{where } K_0 = K \text{ and maps } f_i : \Delta \to K_i$$

with

$$f_i(\Delta) = K_i, \quad f_0 = f$$

and inclusions

$$K_{i+1} \subset L_{i+1}$$

so that $f_{i+1}$ is a lifting of $f_i$ (Figure 2.1). Then for $n$ sufficiently large, the situation is stable, that is, $p_n$ is a homeomorphism.

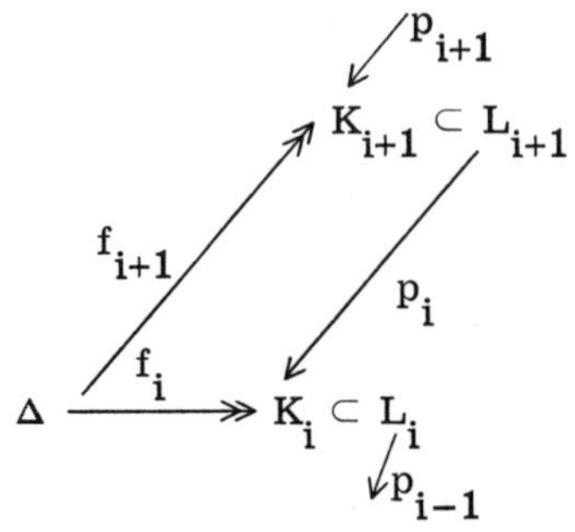


Fig. 2.1

<u>Proof</u>: We can triangulate the whole situation so that $p_i$, $L_{i+1}$, $K_{i+1}$, $f_{i+1}$ are all simplicial. Define the complexity of $f_i$ to be the number of simplexes of $\Delta$ minus the number of simplexes of $K_i$; this number is always nonnegative, and a nontrivial $p_i$ always makes the complexity of $f_{i+1}$ less than that of $f_i$.

2.A.4 Let $p : V' \to V$ be a 2-sheeted covering space of compact 3-manifolds, and let $B' \subset \partial V'$, $B \subset \partial V$ be 2-manifolds, such that $p(B') \subset B$. Let $N$ be a normal subgroup of $\pi_1(B)$, $N' = (p \mid B)_*^{-1}(N)$ which is normal in $\pi_1(B')$. Let $\Delta'$ be a 2-cell in $V'$, with $\partial\Delta' \subset B'$, representing an element of $\pi_1(B')$ not in $N'$. Then there exists a 2-cell $\Delta$ in $V$, with $\partial\Delta \subset B$, representing an element of $\pi_1(B)$ not in $N$.

<u>Proof</u>: We smooth out $p|\Delta'$ so that the singularities of $p(\Delta')$ are only double curves; this is possible since $p|\Delta'$ is an immersion which is at most two-to-one. The singular curves are then of four sorts. Pictured in Figure 2.2 are the inverse images on $\Delta'$, with points mapping to the same point of $p(\Delta')$ labeled the same.

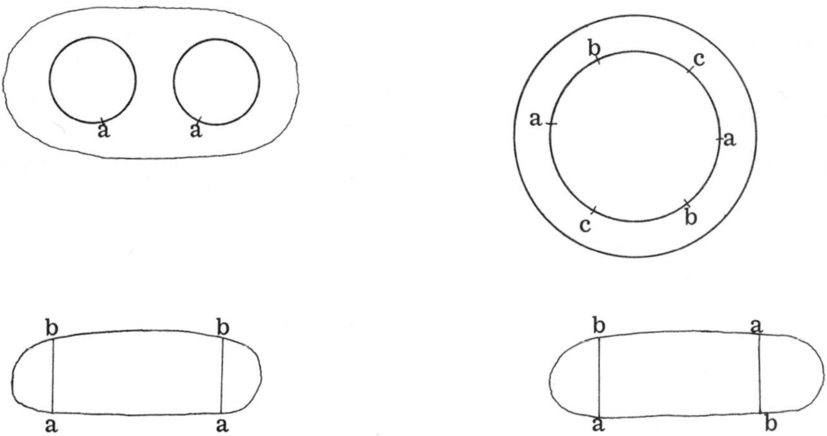


Fig. 2.2

In each of these cases we can change $\Delta'$ and $p|\Delta'$ by means of a "cut" so as to retain the properties (a) $p(\partial\Delta')$ represents an element of $\pi_1(B)$ not in N and (b) the singularities of $p(\Delta')$ are double curves. The number of double curves will be definitely reduced by this modification. In this way we eventually get a nonsingular $\Delta$.

2.A.5 Loop Theorem—Dehn's Lemma: Let M be a 3-manifold, B a 2-manifold $\subset \partial M$, N a normal subgroup of $\pi_1(B)$, such that

$$(\pi_1(B) - N) \cap \ker (\pi_1(B) \to \pi_1(M)) \neq \emptyset .$$

Then there exists a 2-cell $\Delta \subset M$ such that $\partial\Delta \subset B$, and $\partial\Delta$ represents an element of $\pi_1(B)$ not in N.

Proof: We start with a map $f : \Delta \to M$ with $f(\partial\Delta)$ representing an element of $\pi_1(B) - N$. We make f simplicial and let V be a regular neighborhood of $f(\Delta)$ in M. We then construct a tower of 2-sheeted coverings of $f(\Delta)$ and hence of $V : V_0 = v, f_0 = f, B_0 = B \cap V_0, N_0 = \{\alpha \in \pi_1(B_0),$ mapping to N in $\pi_1(B)\}$. Let $p_i : W_{i+1} \to V_i$ be a 2-sheeted covering of $V_i$;

$f_{i+1}$ is a lifting of $f_i$; $V_{i+1}$ is a regular neighborhood of $f_{i+1}(\Delta)$; $B_{i+1} = V_{i+1} \cap p_i^{-1}(B_i)$; $N_{i+1} = \{\alpha \in \pi_1(B_{i+1})$ mapping to $N_i$ in $\pi_1(B_i)\}$.

The tower must terminate, by 2.A.3. At the highest story of the tower, by 2.A.2, a nice, nonsingular 2-cell can be found. By means of 2.A.4, the nonsingular 2-cell with good properties can be made to descend through the tower until it gets to the ground floor, which is the conclusion to be proved.

## 2.B Kneser's Lemma and Other Applications

Let $T$ be a 2-sided (i.e., bicollared) compact 2-manifold in a 3-manifold $M$. Suppose that $\Delta$ is a 2-cell in $M$ such that $\Delta \cap T = \partial\Delta$ is not contractible on $T$. We can adjust it so that $\Delta$ is contained in the interior of $M$ and then find a subset of $M$ homeomorphic to $\Delta \times (-1,+1)$, containing $\Delta$ as $\Delta \times 0$, whose intersection with $T$ is $\partial\Delta \times (-1,+1)$. Let

$$T' = [T - (\partial\Delta \times (-\tfrac{1}{2},+\tfrac{1}{2}))] \cup [\Delta \times \{-\tfrac{1}{2},+\tfrac{1}{2}\}].$$

Then $T'$ is called the reduction of $T$ along $\Delta$; it is the result of a spherical modification of $T$ along $\Delta$ (or surgery of $T$ along $\partial\Delta$, performed within $M$). To any compact 2-manifold $T$ we can assign a complexity:

$$\Sigma(2 - \chi(T_i))^2,$$

$\chi(T_i)$ being the Euler characteristic of $T_i$ summed over all components $T_i$ of $T$. It is easily proved that the complexity of $T'$ is less than that of $T$, and hence the "reduction" terminology is valid.

2.B.1 Kneser's Lemma: Let $T$ be a compact 2-sided 2-manifold in the 3-manifold $M$. Then, after a finite series of reductions at 2-cells $\Delta$, we obtain a 2-sided $T'$ which cannot be further reduced. In this case, if $T_i'$ is any component of $T'$, the homomorphism $\pi_1(T_i') \to \pi_1(M)$ is injective.

Proof: Let $f : \partial\Delta \to T_i'$ be a map representing an element in the kernel of $\pi_1(T_i') \to \pi_1(M)$. Then $f$ extends to a map $f : \Delta \to M$; using the 2-sidedness of $T'$, we can get $f^{-1}(T')$ to consist of a finite number of simple closed curves in the interior of $\Delta$, together with $\partial\Delta$. Let $C$ be an innermost curve in $f^{-1}(T')$, bounding a 2-cell $D$ on $\Delta$, with $D \cap f^{-1}(T') = C$; then $f(C) \subset T'_j$ some component of $T'$. If $f|C$ were not contractible on $T'_j$, then the Loop Theorem and Dehn's Lemma could be applied to $M$ split along $T'$ to find a 2-cell along which $T'$ could be reduced. Since, therefore, $f|C$ is contractible on $T'_j$ we can redefine $f$ to map $D$ into $T'_j$ and to agree with $f$ on $\Delta - D$. Then we can move the resulting map slightly to get a new map $g$ with $g^{-1}(T') = f^{-1}(T') - C$.

Repeating the argument, then, we finally get a map of $\Delta$ into $T'_i$, when $\partial\Delta$ has become the innermost curve in $f^{-1}(T')$, extending the original $f|\partial\Delta$. Thus an arbitrary element in kernel $\pi_1(T'_i) \to \pi_1(M)$ is itself trivial.

The following useful consequence of Kneser's Lemma is also to be found, or something similar to it, in a forthcoming paper by Feustel.

2.B.2 Let $X$ be a topological space containing a bicollared subset $A$, such that if $\tilde{X}$ denotes $X$ split along $A$, and $A_1$, $A_2$ the two copies of $A$ in $\tilde{X}$, then for all base points, $\pi_2(\tilde{X}, A_1 \cup A_2) = 0$. Let $f : M \to X$ be a continuous function, where $M$ is a compact 3-manifold. Then $f$ is homotopic to $g : M \to X$, such that $g^{-1}(A)$ is a reduced 2-sided 2-manifold $T$ in $M$, and $g$ is compatible with the bicollarings of $T$ and $A$. In this situation, if $f_* : \pi_1(M) \to \pi_1(X)$ is injective and if $T_i$ is any component of $T$, then $g_* : \pi_1(T_i) \to \pi_1(A)$ is injective.

Proof: The map $f$ can be smoothed out so that $f^{-1}(A)$ is a bicollared 2-manifold. We apply a reduction to $f^{-1}(A)$ along a 2-cell $\Delta$; this reduction can be gotten via a homotopy of $f$, using the hypothesis $\pi_2(\tilde{X}, A_1 \cup A_2) = 0$ and using homotopy extension properties of subpolyhedra of $M$. Thus we can get $g^{-1}(A)$, where $g$ is homotopic to $f$, to be a bicollared 2-manifold $T$ to which no further reductions apply.

Kneser's Lemma then implies that $\pi_1(T_i) \to \pi_1(M)$ is injective. Thus it follows that, if $\pi_1(M) \to \pi_1(X)$ is injective, since we can factor $\pi_1(T_i) \to \pi_1(X)$ through $\pi_1(A)$, then $\pi_1(T_i) \to \pi_1(A)$ must be injective too.

2.B.3 Kneser's Conjecture: Let M be a compact, conjected 3-manifold, with $\partial M = \emptyset$ ; let $\phi : \pi_1(M) \to A * B$ be a homomorphism <u>onto</u> a free product such that, whenever T is a 2-sided 2-manifold in M, if $\pi_1(T) \to \pi_1(M)$ is injective and $\phi(\pi_1(T)) = \{1\}$, then T is a 2-sphere. Then it is possible to write $M = M_A \cup M_B$ where $M_A \cap M_B$ is a 2-sphere and $\phi(\pi_1(M_A)) = A$, $\phi(\pi_1(M_B)) = B$.

<u>Proof</u>: We concoct aspherical spaces $K_A$ and $K_B$ having fundamental groups A and B and join them along a line segment with middle point p. The resulting space X contains bicollared subset $\{p\}$, and there is a map $f : M \to X$ inducing the given homomorphism $\phi : \pi_1(M) \to A * B \approx \pi_1(X)$.

We now apply 2.B.2 to the situation, being careful, if we wish, that the reductions and homotopies avoid moving the base point of M. We get then f homotopic to g, where $g^{-1}(p)$ is a reduced 2-manifold T which, because of our hypotheses, has to consist of 2-spheres.

Now we use the hypothesis that $\phi$ is onto, to do 1-dimensional surgery on T. If T has more than one component, there is some path $\lambda$ in M whose endpoints lie on different components of T; $g(\lambda)$ represents some element of $A * B$, and so, if $\gamma$ is a loop in M based at the initial point of $\lambda$, mapping into $[g(\lambda)]^{-1}$ in $A * B$ (such $\gamma$ exists since $\phi$ is onto), the path $\gamma\lambda = \mu$ has $g(\mu)$ a contractible loop in X, and $\mu$ joins distinct components of T. By smoothing it, $\mu$ can be written as a product of paths: $\mu = \alpha_1\alpha_2 \cdots \alpha_n$, where only the endpoints of $\alpha_i$ lie on T. Each $g(\alpha_i)$ represents either an element of A or of B. We can reduce the length n of $\mu$ by (1) gluing together $\alpha_i$ and $\alpha_{i+1}$ if both map into A or both into B, (2) omitting $\alpha_i$ if it has both endpoints on the same component of T and has $g(\alpha_i)$ contractible. Since $\mu$ has endpoints on different components of T, the expression for $\mu$ never reduces to length 0. Finally, when the expression for $\mu$ can no longer be reduced, we have either $g(\mu) = g(\alpha_1) \cdots g(\alpha_n)$ is a reduced word in $A * B$, which is impossible since $g(\mu) = 1$ and $n \geq 1$, or else some $g(\alpha_i)$ is contractible in X. This $\alpha_i$ must join distinct components of T and map into $X - \{p\}$ except at the endpoints.

Now we approximate $\alpha_i$ by a nonsingular path which stays away from the base point of M; we use the fact that $g(\alpha_i)$ is contractible in X split along $\{p\}$ and homotopy extension in M to get a function $g' : M \to X$, homotopic to g, such that $g'^{-1}(p)$ consists of $g^{-1}(p)$ except that two components of $g^{-1}(p)$ have been joined by a tube. This process reduces the number of components of $g^{-1}(p)$ so that eventually we get $h : M \to X$ homotopic to g and hence to f, such that $h^{-1}(p)$ is the connected sum of all the components of $g^{-1}(p)$ and hence is a 2-sphere.

We now write $M_A$ and $M_B$ as $h^{-1}$ of the two halves of X and read off the conclusion of the theorem.

## 3 COMBINATORIAL GROUP THEORY

### 3.A A Generalization of the Notion of Amalgamated Free Product of Groups (Pregroups and Their Universal Groups)

In [25], we studied what happens in the situation where a group G has a subset P such that each element of G is representable uniquely by a reduced word in P. It happens that such a G is very similar to a free product.

What happens when the representation by a reduced word is unique only modulo the kind of equivalence that comes up in the theory of amalgamated free products? In this section, we determine the internal structure of the subset P (which we call "pregroups") and prove, following the method of van der Waerden, that its universal group has the desired property. Many interesting examples can be found; they all seem somewhat like amalgamated free products; but there is no simple way of forming all of them out of ordinary amalgamated products. Baer [32] would describe a pregroup as an "add" satisfying his Postulates I-VII and Associative Law T.

#### 3.A.1 Definition and Statement of the Theorem

3.A.1.1 Definition A pregroup consists of:

(a) A set P.

(b) An element of P, denoted by 1.

(c) A function $P \to P$, denoted by $x \mapsto x^{-1}$.

(d) A subset D of $P \times P$.

(e) A function $D \to P$, denoted by $(x, y) \mapsto xy$.

Such that the five following axioms are true:

(1) For all $x \in P$ we have $(1,x)$, $(x,1) \in D$ and $1x = x1 = x$.

(2) For all $x \in P$ we have $(x,x^{-1})$, $(x^{-1},x) \in D$ and $xx^{-1} = x^{-1}x = 1$.

(3) For all $x,y \in P$, if $(x,y) \in D$, then $(y^{-1}, x^{-1}) \in D$ and $(xy)^{-1} = y^{-1}x^{-1}$.

(4) For all $x,y,z \in P$, if $(x,y)$, $(y,z) \in D$, then: $(x,yz) \in D$ if and only if $(xy,z) \in D$, in which case $x(yz) = (xy)z$.

(5) For all $w,x,y,z \in P$, if $(w,x)$, $(x,y)$, $(y,z) \in D$, then either $(w,xy) \in D$ or $(xy,z) \in D$.

We shall often say that $xy$ is defined, instead of $(x,y) \in D$.

3.A.1.2 Definition Let P be a pregroup. A word in P is an n-tuple, for some $n \geq 1$, of elements of P, thus: $(x_1, \ldots, x_n)$. The number n is called the length of the word. It is possible to reduce the word $(x_1, \ldots, x_n)$, if, for some i, we have $x_ix_{i+1}$ defined; then

$$(x_1, \ldots, x_{i-1}, x_ix_{i+1}, x_{i+2}, \ldots, x_n)$$

is called one of its reductions. The word is said to be reduced if no reduction exists; i.e., for all i we have $(x_i, x_{i+1}) \notin D$. Every word of length one is reduced.

If $(x_1, \ldots, x_n)$, $(a_1, \ldots, a_{n-1})$ are words and (where $a_0 = a_n = 1$) the products $x_ia_i$, $a_{i-1}^{-1}x_i$, $a_{i-1}^{-1}x_ia_i$ are all defined, then we define the interleaving of the first by the second to be:

$$(x_1, \ldots, x_n) * (a_1, \ldots, a_{n-1}) = (y_1, \ldots, y_n)$$

where $y_i = a_{i-1}^{-1}x_ia_i$.

We shall prove:

(1) If X is reduced and the interleaving $X * A$ is defined, then $X * A$ is reduced.

(2) The relation on reduced words $X \approx X * A$ is an equivalence relation.

For $a \in P$ and $X = (x_1, \ldots, x_n)$ a reduced word, define $\lambda_a(X)$ thus: If $ax_1$ is not defined,

$$\lambda_a(X) = (a, x_1, \ldots, x_n).$$

If $ax_1$ is defined but $(ax_1)x_2$ not defined,

$$\lambda_a(X) = (ax_1, x_2, \ldots, x_n).$$

If both $ax_1$ and $(ax_1)x_2$ are defined,

$$\lambda_a(X) = ((ax_1)x_2, x_3, \ldots, x_n).$$

(3) If $X$ is reduced, so is $\lambda_a(X)$.

(4) If $X$ is reduced and $ab$ is defined, then

$$\lambda_{ab}(X) \approx \lambda_a(\lambda_b(X))$$

where $\approx$ is defined in (2).

(5) The function $\lambda_a$ induces a function on the set of equivalence classes of reduced words.

(6) The pregroup $P$ can be incorporated into a universal group $U(P)$ such that each element $g \in U(P)$ can be written as a product

$$g = x_1 x_2 \ldots x_n$$

where $(x_1, \ldots, x_n)$ is a reduced word in $P$, and two such reduced words for the same $g$ are $\approx$ equivalent.

Theorem (6) is the main one; the others are lemmas for it, after the proof of van der Waerden [11] for a like result in the context of free products. It has the corollary that every pregroup is contained faithfully in its universal group.

There are many examples of pregroups, of which one gives the free product with amalgamation, another gives what we shall call $A_F \mho \phi$. And there are others more peculiar. Before giving such examples, we shall prove the theorem.

3.A.2 Lemmas (1) (2) (3)

In this section and the next, let $P$ be a fixed pregroup.

3.A.2.1 $(x^{-1})^{-1} = x$.

Proof: Apply axioms (4), (2), and (1) to the product $xx^{-1}(x^{-1})^{-1}$.

3.A.2.2 If $ax$ is defined, then $a^{-1}(ax)$ is defined, and $a^{-1}(ax) = x$. Dually, if $xa$ is defined, so is $(xa)a^{-1}$, and $(xa)a^{-1} = x$.

Proof: By axiom (2), we have $a^{-1}a$ is defined and $= 1$. Thus, by axioms (4) and (1), we have $a^{-1}(ax)$ defined and $= (a^{-1}a)x = x$. The dual case is proved similarly.

3.A.2.3 If $xa$ and $a^{-1}y$ are defined, then: $xy$ is defined if and only if $(xa)(a^{-1}y)$ is defined, in which case $xy = (xa)(a^{-1}y)$.

Proof: Apply axiom (4) and 3.A.2.2 to the product of $x, a, (a^{-1}y)$.

3.A.2.4 If $xa$ and $a^{-1}y$ are defined, then $(x,y,z)$ is a reduced word if and only if $(xa, a^{-1}y, z)$ is reduced. Dually, $(z,x,y)$ is reduced if and only if $(z, xa, a^{-1}y)$ is reduced.

Proof: We must show that if $(x,y,z)$ is reduced, then $(a^{-1}y)z$ is not defined. Suppose that $(a^{-1}y)z$ is defined and consider $\{x, a, a^{-1}y, z\}$. Then, by axiom (5) (and 3.A.2.2 and 3.A.2.1, to prove that $a(a^{-1}y)$ is defined), either $x(a(a^{-1}y))$ is defined or $(a(a^{-1}y))z$ is defined. Since $a(a^{-1}y) = y$, in both cases $(x,y,z)$ is not reduced. Thus, $(a^{-1}y)z$ is not defined if $(x,y,z)$ is reduced; by 3.A.2.3, we have $(xa)(a^{-1}y)$ also not defined; and so $(xa, a^{-1}y, z)$ is reduced.

The converse and the dual are proved in the same way.

It can be shown that axioms (1) through (4) with 3.A.2.4 imply axiom (5). Thus, since axioms (1) through (4) are reasonable and natural and we want to define equivalence classes of reduced words by using 3.A.2.4, we must have axiom (5) for our investigation.

3.A.2.5 If $(x,y)$ is a reduced word and if $xa, a^{-1}y, yb$ are defined, then $(a^{-1}y)b$ is defined.

Proof: If not, by 3.A.2.3 we have $(xa, a^{-1}y, b)$ reduced. Thus, by 3.A.2.4, we have $(x,y,b)$ reduced, in contradiction to having $yb$ defined.

3.A.2.6 If $(x,y)$ is a reduced word and if $xa, a^{-1}y, (xa)b, b^{-1}(a^{-1}y)$ are defined, then $ab$ is defined.

Proof: By 3.A.2.3 twice, $((xa)b)(b^{-1}(a^{-1}y))$ is not defined. Apply axiom (5) to $\{x^{-1}, xa, b, b^{-1}(a^{-1}y)\}$; the consecutive products are defined by 3.A.2.2; the product of the last triple is not defined. Thus by axiom (5) the product of the first triple is defined. By axiom (4) we have $x^{-1}((xa)b) = (x^{-1}(xa))b = ab$, by 3.A.2.2, is defined.

3.A.2.7 Lemma (1): Let $X = (x_1, \ldots, x_n)$ be a reduced word and $A = (a_1, \ldots, a_{n-1})$ a word. Let $a_0 = a_n = 1$. Suppose $x_i a_i$ and $a_{i-1}^{-1}x_i$ are defined. Then $(a_{i-1}^{-1}x_i)a_i$ is defined, and $Y = X * A = (x_1a_1, a_1^{-1}x_2a_2, \ldots, a_{n-1}^{-1}x_n)$ is reduced.

Proof: Apply 3.A.2.4 and 3.A.2.5 to subwords of $X$; 3.A.2.5 shows that $(a_{i-1}^{-1}x_i)a_i$ is defined. By virtue of axiom (4), we can omit the parentheses. Theorem 3.A.2.4 shows that $X * A$ is reduced.

3.A.2.8 Definition By $R_n$ or $R_n(P)$, we denote the set of reduced words in $P$ of length $n$. By $P^{n-1}$, we denote the set of all words of $P$ of length $n-1$. If $A = (a_1, \ldots, a_{n-1})$ and $B = (b_1, \ldots, b_{n-1}) \in P^{n-1}$, and if $a_ib_i$ is defined for all $i$, then we denote by $AB$ the word $(a_1b_1, \ldots, a_{n-1}b_{n-1})$.

3.A.2.9 If $X \in R_n$, $A, B \in P^{n-1}$, and if $X * A$ and $(X * A) * B$ can be defined, then $AB$ can be defined, and then

$$(X * A) * B = X * (AB).$$

Proof: Apply 3.A.2.6 to subwords of $X$. This shows that $AB$ can be defined. Axioms (4) and (3) show that $(X * A) * B = X * (AB)$.

3.A.2.10 Definition The relation $\approx$ on $R_n$ is defined thus:

$$(x_1, \ldots, x_n) \approx (y_1, \ldots, y_n)$$

if and only if there exists $(a_1, \ldots, a_{n-1}) \in P^{n-1}$ such that each $x_ia_i$ and $a_{i-1}^{-1}x_i$ are defined, and $y_i = a_{i-1}^{-1}x_ia_i$. I.e., $X \approx Y$ if and only if there exists $A$ such that $Y = X * A$.

3.A.2.11 Lemma (2): The relation $\approx$ is an equivalence relation on R.

Proof: If $I = (1, \ldots, 1)$, then $X = X * I$, so $X \approx X$. If $A = (a_1, \ldots, a_{n-1})$, let $A^{-1} = (a_1^{-1}, \ldots, a_{n-1}^{-1})$; then, $Y = X * A$ if and only if $X = Y * A^{-1}$; thus, if $X \approx Y$, then $Y \approx X$. If $Y = X * A$ and $Z = Y * B$, then by 3.A.2.9, AB is definable and $Z = X * (AB)$; thus, if $X \approx Y$ and $Y \approx Z$, then $X \approx Z$.

3.A.2.12 Definition By R or R(P) we denote the union of all $R_n$, for $n = 1, 2, \ldots$. For each $a \in P$ and each $X \in R$, we define a word $\lambda_a(X)$ as follows: Let

$$X = (x_1, x_2, x_3, \ldots).$$

(1) If $(a, x_1)$ is reduced, then

$$\lambda_a(X) = (a, x_1, x_2, \ldots).$$

(2) If $ax_1$ is defined but $(ax_1, x_2)$ reduced, then

$$\lambda_a(X) = (ax_1, x_2, x_3, \ldots).$$

(3) If $ax_1$ and $(ax_1)x_2$ are defined, then

$$\lambda_a(X) = ((ax_1)x_2, x_3, \ldots).$$

In case (2) we include, as a degenerate case, the possibility that $ax_1$ is defined and that X has length one.

3.A.2.13 Lemma (3): If X is reduced, then $\lambda_a(X)$ is reduced.

Proof: This is obvious in cases (1) and (2). In case (3), where $ax_1$, $(ax_1)x_2$ is defined, but $x_1x_2$ and $x_2x_3$ are not defined, we must prove that $((ax_1)x_2)x_3$ is not defined. Consider $\{x_1, x_1^{-1}a^{-1}, (ax_1)x_2, x_3\}$ and apply axiom (5) to it; this is possible if $((ax_1)x_2)x_3$ is defined; but axiom (5) implies then that either $x_1x_2$ or $x_2x_3$ is defined, in contradiction to having X reduced.

3.A.3 Lemma (4)

Here we prove

3.A.3.1 Lemma (4): If X is reduced and ab is defined, then $\lambda_{ab}(X) \approx \lambda_a(\lambda_b(X))$.

The proof consists in looking at the various cases. Let $X = (x_1, \ldots, x_n)$.

Case 1 $bx_1$ is not defined. Then,

$$\lambda_b(X) = (b, x_1, \ldots, x_n).$$

Subcase $1_1$ $(ab)x_1$ is not defined. To apply $\lambda_a$, we find ourselves in case 3.A.2.12(2); thus

$$\lambda_a(\lambda_b(X)) = (ab, x_1, \ldots, x_n) = \lambda_{ab}(X).$$

Subcase $1_2$ $(ab)x_1$ is defined. Then, to apply $\lambda_a$ to $\lambda_b(X)$, we are in case 3.A.2.12(3); thus

$\lambda_a(\lambda_b(X)) = ((ab)x_1, x_2, \ldots, x_n)$. It follows from 3.A.2.13 that $((ab)x_1)x_2$ is not defined; so, to apply $\lambda_{ab}$ to X, we are in case 3.A.2.12(2); thus

$$\lambda_{ab}(X) = ((ab)x_1, x_2, \ldots, x_n) = \lambda_a(\lambda_b(X)).$$

Case 2 $bx_1$ is defined but $(bx_1)x_2$ is not defined. Then,

$$\lambda_b(X) = (bx_1, x_2, \ldots, x_n).$$

Subcase $2_1$ $a(bx_1)$ is not defined. Then, $(a, bx_1)$ is reduced; thus $(ab, b^{-1}(bx_1)) = (ab, x_1)$ is reduced by 3.A.2.3. In this case,

$$\lambda_a(\lambda_b(X)) = (a, bx_1, x_2, \ldots, x_n).$$

$$\lambda_{ab}(X) = (ab, x_1, x_2, \ldots, x_n).$$

Thus, $\lambda_{ab}(X) = (\lambda_a(\lambda_b(X))) * (b, 1, \ldots, 1)$.

Subcase $2_2$ $a(bx_1)$ is defined. Then, also is $(ab)x_1$ defined and $= a(bx_1)$.

If $(abx_1)x_2$ is not defined, then

$$\lambda_{ab}(X) = \lambda_a(\lambda_b(X)) = (abx_1, x_2, \ldots, x_n).$$

If $(abx_1)x_2$ is defined, then

$$\lambda_{ab}(X) = \lambda_a(\lambda_b(X)) = ((abx_1)x_2, x_3, \ldots, x_n).$$

Case 3 $bx_1$ and $(bx_1)x_2$ are defined. Then,

$$\lambda_b(X) = ((bx_1)x_2, x_3, \ldots, x_n).$$

Subcase $3_1$ $a(bx_1)$ is not defined. This subcase never occurs because, if it did, then $(ab)x_1$ would be not defined, and

$$\lambda_{ab}(x_1, \ldots, x_n) = (ab, x_1, \ldots, x_n) \approx (a, bx_1, x_2, \ldots, x_n).$$

Thus, using 3.A.2.4, $(bx_1)x_2$ would not be defined, in contradiction to the assumption for case 3.

Subcase $3_2$ $a(bx_1)$ is defined. Then $(ab)x_1$ is defined and $= a(bx_1)$.

Subsubcase $3_{2_1}$ $(abx_1)x_2$ is not defined. Then $a((bx_1)x_2)$ is not defined. Thus:

$$\lambda_{ab}(X) = (abx_1, x_2, \ldots, x_n).$$

$$\lambda_a(\lambda_b(X)) = (a, (bx_1)x_2, x_3, \ldots, x_n).$$

And so

$$\lambda_{ab}(X) = (\lambda_a(\lambda_b(X))) * (bx_1, 1, \ldots, 1).$$

Subsubcase $3_{2_2}$ $(abx_1)x_2$ is defined. Then $(a(bx_1))x_2 = a((bx_1)x_2)$ is defined, and

$$\lambda_{ab}(X) = ((abx_1)x_2, x_3, \ldots, x_n).$$

Hence $((abx_1)x_2)x_3$ is not defined; and so

$$\lambda_a(\lambda_b(X)) = ((abx_1)x_2, x_3, \ldots, x_n) = \lambda_{ab}(X).$$

This exhausts all the possible cases, and so proves 3.A.3.1.

3.A.4 Lemma (5) and the main theorem

For a pregroup P, we have R(P), the set of all reduced words in P, on which we have the equivalence relation $\approx$. By $\widetilde{R}(P)$, we denote the set of $\approx$-equivalence classes.

3.A.4.1 Lemma (5): For all $a \in P$, the function $\lambda_a : R(P) \to R(P)$ induces a function, also denoted $\lambda_a : \widetilde{R}(P) \to \widetilde{R}(P)$.

Proof: We must show that, if $Y = X * B$, then $\lambda_a(X) \approx \lambda_a(Y)$. Let $X = (x_1, \ldots, x_n)$ and $B = (b_1, \ldots, b_{n-1})$. There are three cases as follows:

(1) $ax_1$ is not defined. Let $B' = (1, b_1, \ldots, b_{n-1})$. In this case,

$$\lambda_a(Y) = (\lambda_a(X)) * B'.$$

(2) $ax_1$ is defined but $(ax_1)x_2$ not defined. In this case,

$$\lambda_a(Y) = (\lambda_a(X)) * B.$$

(3) $ax_1$ and $(ax_1)x_2$ are both defined. Let

$$B'' = (b_2, \ldots, b_{n-1}).$$

In this case,

$$\lambda_a(Y) = (\lambda_a(X)) * B''.$$

In each case, the word $\lambda_a(X)$ is determined; the expression on the right-hand side of these formulas determines a reduced word, the examination of which determines $\lambda_a(Y)$. For example, in the most difficult case, case (3),

$$\lambda_a(X) = ((ax_1)x_2, x_3, \ldots, x_n).$$

The triple product $(ax_1)x_2b_2$ is defined by the dual of 3.A.2.5 since $(x_2, x_3)$ is reduced, $x_2b_2$ and $b_2^{-1}x_3$ is defined, and $(ax_1)x_2$ is defined. Similarly, $ax_1b_1$ is defined. Hence,

$$(ax_1b_1)b_1^{-1}x_2b_2 = (ay_1)y_2$$

is defined by 3.A.2.3, and so we have:

$$\lambda_a(X) * B'' = ((ax_1)x_2b_2, b_2^{-1}x_3b_3, \ldots, b_{n-1}^{-1}x_n)$$

$$= ((ax_1b_1)b_1^{-1}x_2b_2, \ldots, b_{n-1}^{-1}x_n)$$

$$= ((ay_1)y_2, \ldots, y_n) = \lambda_a(Y).$$

3.A.4.2 Definition Let P, Q be two pregroups. A function $\phi : P \to Q$ is called a morphism of pregroups if, for all $x, y \in P$ such that $xy$ is defined, we have $\phi(x)\phi(y)$ defined, and $\phi(xy) = \phi(x)\phi(y)$.

The class of pregroups with their morphisms constitutes a category of groups and homomorphisms. There exists, by abstract nonsense (the Adjoint Functor Theorem), a functor coadjoint to the functor of inclusion. This gives the universal group U(P) of a pregroup, i.e., U(P) is a group, and there is a specific morphism $\iota : P \to U(P)$, such that, for every group G and every morphism $\phi : P \to G$, there exists a unique homomorphism $\psi : U(P) \to G$ such that $\phi = \psi \circ \iota$.

3.A.4.3 Let S be the group of permutations of $\widetilde{R}(P)$. Then $\lambda$ is a morphism of P into S.

<u>Proof</u>: Since $\lambda_1$ is the identity function of $\widetilde{R}(P)$ into itself, and $\lambda_x \circ \lambda_{x^{-1}} = \lambda_{x^{-1}} \circ \lambda_x = \lambda_1$, by 3.A.3.1, every $\lambda_x$ belongs to S. Thus, 3.A.3.1 may be interpreted as saying that $\lambda$ is a morphism.

3.A.4.4 By the universal property, $\lambda$ extends uniquely to a homomorphism, also denoted $\lambda$, of U(P) into S. We denote the value of $\lambda$ on $g \in U(P)$, by $\lambda_g : \widetilde{R}(P) \to \widetilde{R}(P)$.

Since $\iota(P)$ generates U(P), each $g \in U(P)$ can be written as $g = \iota(x_1)\iota(x_2) \ldots \iota(x_n)$ where $(x_1, x_2, \ldots, x_n)$ is a word in P. After applying reductions to this word, we obtain a reduced word $(x_1, \ldots, x_n)$ such that

$$g = \iota(x_1)\iota(x_2) \ldots \iota(x_n).$$

We denote by $\Lambda$, the word (1) of length one. We have the formula:

$$\lambda_g([\Lambda]) = \lambda_{x_1}(\lambda_{x_2}(\ldots(\lambda_{x_n}([\Lambda]))\ldots)) = [(x_1, x_2, \ldots, x_n)].$$

Here, $[\cdot]$ denotes the $\approx$-equivalence class. Each application of a

$$\lambda_{x_i}$$

here is in the case 3.A.2.12(1).

In this manner, g determines by itself the class of reduced words which represent g. Thus:

3.A.4.5 Theorem: If P is a pregroup, then each element $g \in U(P)$, the universal group of P, can be represented as a product $x_1 x_2 \ldots x_n$ of a reduced word in P, $(x_1, \ldots, x_n)$. Two such reduced words represent the same element of U(P), if an only if they are $\approx$ equivalent. (Here we have identified $x \in P$ with $\iota(x) \in U(P)$ to simplify the notation.)

Proof: We have just proved the "only if." The "if" is the trivial computation that, if $g = x_1 x_2 \ldots x_n$, then $g = (x_1 a_1)(a_1^{-1} x_2 a_2) \ldots (a_{n-1}^{-1} x_n)$.

3.A.4.6 Corollary: A pregroup P is contained faithfully in its universal group U(P).

This means that the specific morphism $\iota : P \to U(P)$ is injective. This follows from the theorem, since no word of length one is equivalent to any other word.

3.A.5 Examples

3.A.5.1 The most standard example of a pregroup is made of three groups A, B, C and of two monomorphisms $\phi : C \to A$, $\psi : C \to B$. Identify $\phi(C)$ with $\psi(C)$; then $A \cap B = C$. Let $P = A \cup B$. The 1 and the inverse are obvious; the product is defined for two elements x, y if and only if the two belong to one or the other of A or B. The axioms (1) through (4) are clearly satisfied. For axiom (5), it breaks into simple cases easy to verify. The universal group is the free product with amalgamation $A *_C B$.

3.A.5.2 Here is a similar case but more general. A tree of groups consists of:

(a) A set I, partially ordered by $<$, with least element, such that for all $i, j, k \in I$, if $i \leq k$ and $j \leq k$, then either $i \leq j$ or $j \leq i$. (Such an ordered set is a sort of abstract tree.)

(b) A class of groups $\{G_i\}$ indexed by $i \in I$.

(c) For all $i, j \in I$, if $i < j$, a monomorphism $\phi_{ij} : G_i \to G_j$.

This structure is to satisfy the condition that, for all $i, j, k \in I$, if $i < j < k$, then

$$\phi_{jk} \circ \phi_{ij} = \phi_{ik} : G_i \to G_k .$$

We can construct, as above, the union P of all $\{G_i\}$, identifying $x \in G_i$ with $\phi_{ij}(x) \in G_j$. The reader can verify that, by virtue of the properties of the tree, with the obvious operations, P is a pregroup. The universal groups of such pregroups include all ordinary free products with amalgamation with many factors.

3.A.5.3 Consider a free amalgamated product $A *_C B$. Let P be the subset of all elements that can be written $bab'$, for some $b, b' \in B$, $a \in A$; thus, P contains A and B and somewhat more. Say that the product xy of two elements $x, y \in P$ is defined, when $xy \in P$. Using the structure (by reduced words in $A \cup B$, etc.) of $A *_C B$, we can prove that P is a pregroup. The universal group of P is again $A *_C B$; but the structures of $A *_C B$ by words in P is different from that by words in $A \cup B$.

3.A.5.4 Consider a group G with subgroup H. Let P be the set G, but define multiplication of x and y if and only if at least one of $\{x, y, xy\}$ belongs to H. This is a pregroup, and its universal group is not too like a free amalgamated product.

3.A.5.5 Let G be a group, H a subgroup, and $\phi : H \to G$ a monomorphism. Construct four sets in 1-to-1 correspondence with G:

$$G, \quad x^{-1}G, \quad Gx, \quad x^{-1}Gx.$$

Identify $h \in H \subset G$, with $x^{-1}\phi(h)x \in x^{-1}Gx$. Define multiplication between G and G, G and Gx, $x^{-1}G$ and G, $x^{-1}G$ and Gx, Gx and $x^{-1}G$, Gx and $x^{-1}Gx$, $x^{-1}Gx$ and $x^{-1}G$, $x^{-1}Gx$ and $x^{-1}Gx$, by cancellation of $xx^{-1}$ and multiplication in G. By the formulas:

$$hx^{-1} = x^{-1}\phi(h), \quad \text{and} \quad xh = \phi(h)x,$$

which follows from identification of H with $x^{-1}\phi(H)x$, multiplication is defined in all cases when one factor belongs to H. This monstrosity is a pregroup, whose universal group we shall call $G_H\mathfrak{I}\phi$.

### 3.B Bipolar Structures and Finite Amalgamated Subgroups

We now consider a kind of combinatorial group-theoretic structure which is more specialized than that considered in the preceding section. It is a generalization of the situation of a free product with finite amalgamated subgroup.

3.B.1 A bipolar structure on a group G is a partition of G into six disjoint sets, termed

F, S, EE, EE*, E*E, E*E*

satisfying the axioms below. We let X, Y, Z be symbols standing for the letters E or E* and suppose that, if X = E or E*, then X* = E* or E, respectively.

Axioms:

1. F is a finite subgroup of G.
2. $F \cup S$ is a subgroup of G in which F has index 1 or 2.
3. If $f \in F$, $g \in XY$, then $gf \in XY$.
4. If $s \in S$, $g \in XY$, then $gs \in XY^*$.
5. If $g \in XY$, then $g^{-1} \in XY$.
6. If $g \in XY$, $h \in Y^*Z$, then $gh \in XZ$.
7. If $g \in G$, there exists N(g) such that whenever $g_1, \ldots, g_n \in G$ and there exist $X_0, X_1, \ldots, X_n$ such that $g_i \in X^*_{i-1}X_i$, and $g = g_1g_2 \ldots g_n$, then $n \leq N(g)$.
8. $EE^* \neq \emptyset$.

Axioms 7 and 8 are to some extent optional in that much can be proved without them. Axiom 7 implies that G is generated by "irreducible" elements, while axiom 8 implies a sort of nontriviality.

The elements of F are, so to speak, the stabilizers of the structure, while those of S, which may be empty, are the involutions.

3.B.2 An element $p \in G$ is said to be irreducible, if either $p \in F \cup S$, or $p \in XY$, such that p cannot be written $p = gh$ for $g \in XZ$, $h \in Z^*Y$. That is, an element p is irreducible if axiom 7 holds with $N(p) = 1$.

3.B.2.1 Every element of G is a product of finitely many irreducible elements.

This follows from axioms 7 and 6.

3.B.2.2 If $g \in XY$, $p \in YZ$, and p is irreducible, then either $gp \in F \cup S$, or $gp \in XW$ for some W.

Proof: The other possibility is $gp \in X^*W$. By axiom 5, $g^{-1} \in YX$. Then, from the definition, since $p = (g^{-1})(gp)$, it cannot be irreducible.

3.B.2.3 Dually, if $g \in XY$, $p \in ZX$, and p is irreducible, then $pg \in WY$ for some W or $pg \in F \cup S$.

3.B.2.4 If $p \in XY$, $q \in YZ$ are both irreducible, then pq is in $F \cup S \cup XZ$ and is irreducible. If $pq \in F \cup S$, then $pq \in F$ when $X = Z$, and $pq \in S$ when $X = Z^*$.

Proof: If $pq \in F$, then, by axioms 5 and 3, $q = p^{-1}(pq) \in YX$, and so $X = Z$. All the other parts of this proposition have been proved in 3.B.2.2 and 3.B.2.3 except for the irreducibility of pq. If pq is not irreducible, then it belongs to XZ and $pq = gh$, for $g \in XW$, $h \in W^*Z$ for some g, h, and W. We then have $p = g(hq^{-1})$, and, by axiom 5 and 3.B.2.3, we have $hq^{-1} \in W^*U$ for some U or $hq^{-1} \in F \cup S$. It cannot happen that $hq^{-1} \in W^*U$ since p is irreducible. If $hq \in F$, then axiom 3 applied to $p = g(hq^{-1})$ and axioms 1, 3, and 5 applied to $q = [h^{-1}(hq^{-1})]^{-1}$, show that $p \in XW$ and $q \in W^*Z$, contradictory to $p \in XY$, $q \in YZ$. On the other hand, if $hq^{-1} \in S$, then we apply axioms 1, 4, and 5 to the same products to see $p \in XW^*$ and $q \in WZ$, which is again a contradiction. Hence pq must be irreducible.

3.B.2.5 If $p \in XY$ and $q \in F \cup S$, p being irreducible, then pq is irreducible.

The proof is very similar to that of 3.B.2.4.

3.B.2.6 If p is irreducible, then so is $p^{-1}$.

3.B.2.7 If a, b, c, ab, bc are irreducible, then abc is irreducible unless $b \in F \cup S$.

Proof: If $b \in XY$, then $ab \in F \cup S$ or $ab \in ZY$, by 3.B.2.4 and 3.B.2.5. If $ab \in F \cup S$, then abc is irreducible, by 3.B.2.5 and 3.B.2.6. If $ab \in ZY$, then if $c \in F \cup S$, then abc is irreducible by 3.B.2.5; while, if $c \notin F \cup S$, then since bc is irreducible, we must have $c \in YW$, in which case we apply 3.B.2.4, having $ab \in ZY$ to show abc is irreducible.

3.B.2.8 If a, b, c, d, ab, bc, cd are irreducible, then either abc or bcd is irreducible.

For, by 3.B.2.7, otherwise both b and c belong to $F \cup S$, so that $bc \in F \cup S$, and then by 3.B.2.5 and 3.B.2.6, both abc and bcd are irreducible.

3.B.3.1 Let P consist of the irreducible elements of G. Let D consist of those ordered pairs $(a, b) \in P \times P$, for which the product in G, $ab \in P$. Then P with 1, inverse, and multiplication as in G, but restricted only to D, is a pregroup.

Proof: This follows easily from the fact that G is a group and from propositions 3.B.2.6 and 3.B.2.8.

We can now speak of reduced words in the pregroup P; it turns out that a word $(p_1, \ldots, p_n)$ is reduced if and only if, when $n > 1$, there exist $X_0, \ldots, X_n$, such that $P_i \in X^*_{i-1}X_i$.

3.B.3.2 Let $(p_1, \ldots, p_n)$ and $(q_1, \ldots, q_m)$ be two reduced words in P such that

$$p_1p_2 \cdots p_n = q_1q_2 \cdots q_m.$$

Then $m = n$ and there exist $c_0 = 1, c_1, \ldots, c_{n-1} \in F \cup S$, $c_n = 1$ such that, for all i, $q_i = c_{i-1}^{-1}p_ic_i$.

Proof: We assume $n \leq m$ and prove the proposition by induction on n.

If $n = 1$, we must have $m = 1$, since $p_1 = q_1 q_2 \ldots q_m$ is irreducible; it also follows that $p_1 = q_1$.

For $n \geq 2$, if we have $p_1 p_2 \ldots p_{n-1} p_n = q_1 q_2 \ldots q_m = g$, then $g \in XY$ for some X and Y, and so $p_n \in UY$, $q_m \in VY$ for some U and V, and so $q_m p_n^{-1} \in VU$ or $F \cup S$, and $q_m p_n^{-1}$ is irreducible, by 3.B.2.4. If $q_m p_n^{-1} \in VU$, then $(q_1, q_2, \ldots, q_{m-1}, q_m p_n^{-1})$ is still a reduced word since $q_{m-1} \in WV^*$; then we have

$$p_1 p_2 \ldots p_{n-1} = q_1 q_2 \ldots q_{m-1}(q_m p_n^{-1}),$$

and so by induction, $n - 1 = m$, in contradiction to the assumption $n \leq m$.

Therefore, $q_m p_n^{-1} \in F \cup S$; then the word $(q_1, q_2, \ldots, q_{m-1}(q_m p_n^{-1}))$ is reduced; and so, by induction, $n - 1 = m - 1$, and there are $c_0 = 1$, $c_1, c_2, \ldots, c_{n-1} \in F \cup S$, such that

$$q_i = c_{i-1}^{-1} p_i c_i \quad \text{for } i < n - 1$$

$$q_{n-1}(q_n p_n^{-1}) = c_{n-2}^{-1} p_{n-1}.$$

By defining $c_{n-1} = (q_n p_n^{-1})^{-1} = p_n q_n^{-1}$, we complete the inductive step.

3.B.3.3 G is the universal group of the pregroup P of the irreducible elements in its bipolar structure.

This follows from 3.B.3.2 and 3.B.2.1

3.B.4 Define, when we have a bipolar structure on a group G,

$G_1 = F \cup \{\text{irreducible elements of } EE\}$

$G_2 = F \cup \{\text{irreducible elements of } E^*E^*\}$

3.B.4.1 $G_1$ and $G_2$ are subgroups of G.

This follows from axioms 1, 3, 5, and from 3.B.2.4 and the observation that, if $x, y \in EE$, then $xy$ is not an element of S; otherwise $y = x^{-1}(xy)$ would belong to $EE^*$, by axioms 5 and 4.

3.B.4.2 If $S \neq \emptyset$, then $G = \{F \cup S\} *_F G_1$.

Proof: Let $s \in S$. Then the irreducible elements of the bipolar structure consist of

$$G_1 \subset F \cup EE \qquad sG_1 \subset S \cup E^*E$$

$$G_1 s \subset S \cup EE^* \qquad sG_1 s = G_2 \subset F \cup E^*E^* .$$

Comparing this with the example 3.A.5.3, we see that the pregroup of irreducible elements is isomorphic with the pregroup of this example, where A is $G_1$ and B is $F \cup S$. Hence the universal group is the free product with amalgamation.

3.B.4.3 If $S = \emptyset$ and there is no irreducible element of $EE^*$, then $G = G_1 *_F G_2$.

Proof: In this case, the pregroup of irreducible elements is simply $G_1 \cup G_2$, and thus we are essentially in the situation of example 3.A.5.1.

3.B.4.4 If $S = \emptyset$, and there is an irreducible element $t \in EE^*$, then $tFt^{-1} \subset G_1$, and, if $\phi : F \to G_1$ is defined to be the function $f \to tft^{-1}$, then $G = G_1 {}_F\mathfrak{Z}\, \phi$.

Proof: If $f \in F$, then $tf$ is an irreducible element of $EE^*$ by axiom 3 and 3.B.2.5, and hence, by 3.B.2.4, $(tf)t^{-1}$ is an irreducible element of $EE$ or an element of $F$ and hence in $G_1$. We can build up the entire pregroup of irreducible elements out of $t$ and $G_1$; if $x \in E^*E$, for example, then $tx \in G_1 \subset F \cup EE$, and so $x = t^{-1}(tx)$; the irreducible elements then consist of:

$$G_1 \subset F \cup EE \qquad t^{-1}G_1 \subset E^*E$$

$$G_1 t \subset EE^* \qquad t^{-1}G_1 t = G_2 \subset F \cup E^*E^*$$

These four sets are all disjoint, except that $G_1 \cap t^{-1}G_1 t = F = t^{-1}(\phi(F))t$.

Comparing with example 3.A.5.5, we see that this pregroup is exactly isomorphic to the one described there, and hence the universal group G can be named as we named it there, $G_1 {}_F\mathfrak{Z}\phi$.

3.B.5 If G has a bipolar structure satisfying axiom 8, the nontriviality axiom, then G can be written in one of two forms:

$$A *_F B \quad \text{or} \quad A_F \mathfrak{I} \phi$$

where F is a finite subgroup of G and where in the first case, F, is properly contained in both A and B.

Proof: This is a summary of 3.B.4. If, say, $F = B$ in the first case, then $G = A$, and, going back to 3.B.4.2 and 3.B.4.4 we see that G would consist only of $F \cup S$ or $F \cup EE$, in both of which cases $EE^* = \emptyset$, contradicting axiom 8.

3.B.6 Topological Remark

Let $K_A$ and $K_B$ be complexes having fundamental groups A and B and intersecting in a connected subcomplex having fundamental group $F = A \cap B$. Then, by Seifert's [5] or van Kampen's [7] Theorems, the fundamental group of $K_A \cup K_B$ is $A *_F B$.

If $K_A$ contains two isomorphic subcomplexes $K_F$, $K'_F$ with fundamental groups F and $\phi(F)$ and we attach to $K_A$ the cylinder $K_F \times [0, 1]$, identifying $K_F$ with $K_F \times 0$ and $K_F$ with $K_F \times 1$, we obtain a complex with fundamental group $A_F \mathfrak{I} \phi$. This too follows from the Seifert-van Kampen Theorem.

Thus the combinatorial situation of bipolar structure is somewhat paralleled by a topological situation. The topological picture is very useful to the intuition. The exact picture of the case when $S \neq \emptyset$ is a bit complicated and involves the attachment of the mapping cylinder of a connected two-sheeted covering space to another space. It might be an interesting research project to investigate what kinds of identification spaces have fundamental groups with natural structures similar to bipolar structures.

# 4 THE THEORY OF ENDS

## 4.A Ends of Groups

The subject can be approached either from a combinatorial or a topological point of view. Here we choose the former.

4.A.1 Let G denote a group. By $\mathcal{A}$ or $\mathcal{A}(G)$ is meant the Boolean algebra of all subsets of G. Then A + B denotes the symmetric difference of A and B; AB denotes the intersection. (There is some chance of confusion here between intersection and the set of all products ab in G. In this section AB always means intersection.) Further, 1 denotes G, the unit element of $\mathcal{A}$, and 0 denotes $\emptyset$; $A^* = 1 + A = G\setminus A$ is the notation for complementation. On $\mathcal{A}$ there are compatible left and right actions of G as automorphisms of $\mathcal{A}$.

4.A.2 For $g \in G$, let $\nabla_g(A) = A + gA$. For this simple operation we can write several rules:

$$\nabla_g(A + B) = \nabla_g(A) + \nabla_g(B)$$

$$\nabla_g(AB) = \nabla_g(A)B + (gA)\nabla_g(B)$$

$$\nabla_g(1) = \nabla_g(0) = 0$$

$$\nabla_g(Ah) = \nabla_g(A)h$$

$$\nabla_g(hA) = h\nabla_{h^{-1}gh}(A)$$

$$\nabla_{g^{-1}}(A) = g^{-1}\nabla_g(A)$$

$$\nabla_{gh}(A) = \nabla_g(A) + g\nabla_h(A)$$

$$\nabla_g(A) = 0 \quad \text{for all} \quad g \Longrightarrow A = 0 \quad \text{or} \quad A = 1$$

4.A.3 Let $\mathcal{F}$ denote the subset of $\mathcal{A}$ consisting of all finite subsets of G. $\mathcal{F}$ is an ideal in $\mathcal{A}$, closed under left- and right-G operators and closed under $\nabla_g$ for all g.

Let $\mathcal{Q}(G)$ denote the set of all those $A \in G$, $\nabla_g(A) \in \mathcal{F}$. It follows from the properties of $\nabla_g$ that $\mathcal{Q}$ is a subalgebra of $\mathcal{A}$, containing $\mathcal{F}$ and that $\mathcal{Q}$ is closed under left- and right-G operators.

Let $\mathcal{E}(G)$ denote $\mathcal{Q}(G)/\mathcal{F}$. Then $\mathcal{E}$ is a Boolean algebra with right G operators; the induced left-G action on $\mathcal{E}$ is trivial; and in fact $\mathcal{E}$ is just the subgroup of $\mathcal{A}/\mathcal{F}$ left fixed by left-G action.

The structure theorem for Boolean algebras states that the maximal ideal space E of $\mathcal{E}$ is a 0-dimensional compact Hausdorff space and that $\mathcal{E}$ can be identified with the class of clopen sets of E. The space E is called the space of ends of G; the number of ends is then (identifying all infinite numbers) the rank of $\mathcal{E}$ as vector space over GF(2).

4.A.4 Let T be any set generating G. It follows from the properties listed in 4.A.2, that $\mathcal{Q}(G)$ can be defined only in terms of T. That is

$$\mathcal{Q}(G) = \{A \in \mathcal{A} \mid \text{for all } t \in T, \nabla_t(A) \in \mathcal{F}\}$$

The graph of G with respect to T, $\Gamma(G, T)$, consists of two sets, $\Gamma_0 = G$ and $\Gamma_1 = T \times G$, and two functions $v_1, v_2 : \Gamma_1 \to \Gamma_2$, defined by $v_1(t, g) = g$, $v_2(t, g) = tg$. Then T generates G if and only if $\Gamma$ is connected. There is a right action of G on the graph $\Gamma$.

We can look at mod 2 cochains on $\Gamma$. The 0-dimensional cochains can be identified with $\mathcal{A}(G)$. Given such a 0-cochain A, its coboundary has the value on an edge (t, g) equal to 0 if g and tg lie both in A or both in A* and has the value equal to 1 if g and tg lie in different sets A and A*. Thus

$$\delta A = \{(t, g) \mid g \in \nabla_{t^{-1}}(A)\}.$$

This shows that when T is a finite set, then $A \in \mathcal{Q}(G)$ if and only if $\delta A$ is finite.

4.A.5 At this point we could say a word about locally finite graphs $\Gamma$ in general. By $\mathcal{Q}(\Gamma)$ is meant the set of those 0-cochains A whose coboundary $\delta A$ is finite; and by $E(\Gamma)$ is meant the Boolean algebra $\mathcal{Q}(\Gamma)$/(finite 0-cochains); and by an end of $\Gamma$ is meant a maximal ideal in $E(\Gamma)$.

Thus, in the case of the graph of a group with respect to some finite generating set, we can identify the ends of the group and the ends of the graph.

4.A.6 We now list more or less classical theorems about the ends of groups, cf. [6], [10]. Here we use the abstract definition, not including any assumption on finite generation unless explicitly stated. We shall then indicate proofs.

4.A.6.1 If N is a finite normal subgroup of G, then $E(G) \approx E(G/N)$.

4.A.6.2 If H is a subgroup of finite index in G, then $E(G) \approx E(H)$.

4.A.6.3 If G contains a finitely generated, infinite, normal subgroup H of infinite index in G, then $E(G) \approx Z_2$, (i.e., G has one end).

4.A.6.4 If there is $A \in E(G)$, $A \neq 0$ and $A \neq 1$, and there are infinitely many $g \in G$ such that $Ag = A$ and G is finitely generated, then G has two ends.

4.A.6.5 If G is a finitely generated group with two ends, then there is a finite normal subgroup N such that G/N is either infinite cyclic or isomorphic to $Z_2 * Z_2$ (= the infinite dihedral group).

4.A.6.6 If G is a free product with finite amalgamated subgroup, $G = A *_F B$, where F is a proper subgroup of both A and B, and of index $\geq 3$ in B, then G has infinitely many ends. Also if $G = A_F \circlearrowleft \phi$, where F is a proper finite subgroup of A, then G has infinitely many ends.

4.A.7 Proofs

Proof of 4.A.6.1 N a finite normal subgroup of G implies $E(G) \approx E(G/N)$.

Let $\phi : G \to G/N$ be the quotient homomorphism. For $A \subset G/N$, define $f(A) = \phi^{-1}(A) \subset G$. Then f induces maps $A(G/N) \to A(G)$, $F(G/N) \to F(G)$,

$Q(G/N) \to Q(G)$, and $E(G/N) \to E(G)$. Let H be a set of coset representatives of G/N in G, containing $1 \in N$. Define, for $B \subset G$, $s(B) = \phi(B \cap H)$; s defines maps on all levels, in particular from $E(G) \to E(G/N)$. The one point which may not be obvious is that if $\nabla_g(B)$ is finite for all $g \in G$, then $\nabla_h(\phi(B \cap H))$ is finite for all $h \in G/N$.

To see this, let $B' = \phi^{-1}(\phi(B))$; thus $B'$ is a union of cosets of N. If we know that $\nabla_g(B')$ is finite for all g, then it is easy to see $\nabla_h(\phi(B' \cap H))$ is finite for all h. Now,

$$\begin{aligned} B' + B &= \phi^{-1}(\phi(B)) \setminus B \\ &= \bigcup_{n \in N} \{nB\} \setminus B \\ &\subset \bigcup_{n \in N} \{nB\} \setminus \bigcap_{n \in N} \{nB\} \\ &= \bigcup_{n \in N} \{(B + nB)\} \\ &= \bigcup_{n \in N} \{(\nabla_n B)\} \end{aligned}$$

This latter is a finite set. In this computation, we used the identity, valid in any Boolean algebra,

$$\bigcup_{i=1}^{k} (A + B_i) = A \cup \bigcup_{i=1}^{k} \{B_i\} \setminus A \cap \bigcap_{i=1}^{k} \{B_i\}.$$

This means now that $B'$ differs from B by a finite set, and so $B' \in Q(G)$ if $B \in Q(G)$.

Having now s from $E(G)$ to $E(G/N)$, we see that sf, i.e., $A \to \phi^{-1}(A) \to \phi(H \cap \phi^{-1}(A))$, is the identity map and that, for sets of the form $B'$ as above, fs, i.e., $B' \to \phi(H \cap B') \to \phi^{-1}(\phi(H \cap B'))$, is the identity. Hence, on the $E$ level, where every element of $E(G)$ is representable by $B'$, f and s are inverses of each other.

Proof of 4.A.6.2 $H \subset G$, $[G : H] < \infty$, implies $E(G) \approx E(H)$.

Lemma: This is true under the additional hypothesis that H is a normal subgroup of G.

Let $A \subset G$; the map $f : \mathcal{A}(G) \to \mathcal{A}(H)$, defined by $f(A) = A \cap H$ induces maps on all levels, in particular from $\mathcal{E}(G)$ to $\mathcal{E}(H)$. Let $\{g_1, \ldots, g_n\}$ be representatives of the cosets Hg with $g_1 = 1$; for $B \subset H$, define $s(B) = g_1 B \cup \ldots \cup g_n B \subset G$. Clearly s induces a map $\mathcal{A}(H) \to \mathcal{A}(G)$ such that $f(s(B)) = B$; it remains to be proved that s induces a map on the $\mathcal{E}$ level and that, for $A \in \mathcal{Q}(G)$, $sf(A) + A$ is finite; this will show that on $\mathcal{E}$, s and f are inverses of each other.

Let $B \in \mathcal{Q}(H)$ so that, for all $h \in H$, $\nabla_h(B)$ is finite; we compute

$$\begin{aligned} \nabla_g(s(B)) &= \nabla_g(g_1 B + \ldots + g_n B) \\ &= \sum_{i=1}^{n} (g_i B + g_{\tau(i)} h_i B) \\ &= \sum_{i=1}^{n} g_{\tau(i)} \nabla_{h_i}(B) \end{aligned}$$

Where $gg_i = g_{\tau(i)} h_i$, $h_i \in H$, and $\tau$ some permutation of $\{1, \ldots, n\}$.

Then $\nabla_g(s(B))$ is finite since $\nabla_{h_i}(B)$ is finite for all $h_i$. This computation shows that s induces a map $\mathcal{E}(H) \to \mathcal{E}(G)$.

Given $A \in \mathcal{Q}(G)$,

$$\begin{aligned} sf(A) + A &= s(A \cap H) + A \\ &= \sum_{i=1}^{n} g_i(A \cap H) + A \cap \sum_{i=1}^{n} g_i H \\ &= \sum_{i=1}^{n} (g_i A + A) \cap (g_i H) \\ &= \sum_{i=1}^{n} \nabla_{g_i}(A) \cap (g_i H) \end{aligned}$$

which is finite. This shows sf induces the identity on $\mathcal{E}(G)$. This proves the lemma.

Now for the proof of Theorem 4.A.6.2 itself: If H is of finite index in G, then let K be the intersection of all conjugates of H. Now K is still of finite index in G; we have in fact $[G : K] \leq [G : H]!$. And K is a normal subgroup of G, from which it follows that K is normal in H as well. By the lemma $E(G) \approx E(K)$ and $E(H) \approx E(K)$; so $E(G) \approx E(H)$.

Proof of 4.A.6.3 If H normal in G, $[G : H] = \infty$, H infinite, finitely generated, then $E(G) = Z_2$.

Let $A \in Q(G)$. It must be shown that A or A* is finite.

Let $T = \{h_1, \ldots, h_n\}$ be a finite set of generators of H. Let S be a set of representatives of the cosets G/H. For $s \in S$, let $A_s = A \cap (Hs)$. Then, for $h_i \in T$ we have

$$\nabla_{h_i}(A) = \bigcup_{s \in S} \nabla_{h_i}(A_s);$$

and since $A \in Q(G)$, we conclude that, for all but finitely many s,

$$\nabla_{h_i}(A_s) = \emptyset.$$

Hence, for all but finitely many s and for all $h_i \in T$, we have

$$\nabla_{h_i}(A_s) = \emptyset;$$

in this latter case, $(A_s)s^{-1} \subset H$ and has for all $h_i \in T$ the formula

$$\nabla_{h_i}(A_s)s^{-1} = 0.$$

It follows that, by the last rule in 4.A.2 and the computations in 4.A.4, $(A_s)s^{-1} = \emptyset$ or $= H$. Thus, for all but finitely many $s \in S$, $A_s = \emptyset$ or $A_s = Hs$; in particular, there exists $s \in S$, such that $A_s = \emptyset$ or $A_s = Hs$.

If $A_s = \emptyset$ for one s and $A_{s'}$ is infinite for another s′, then

$$\nabla_{s's^{-1}}(A) = A + s's^{-1}A$$

intersects s′H in $A'_s + s's^{-1}A_s = A'_s$ and so is infinite, contradicting

$A \in \mathcal{Q}(G)$. Thus, if $A_s = \emptyset$ for some $s$, then $A_s$ is $\emptyset$ for all but finitely many $s$ and is finite for the remaining $s$. Hence $A$ is finite.

Similarly, if $A = Hs$ for some $s$, the above argument, on $A^*$, shows $A^*$ is finite.

Remark to 4.A.6.3 This theorem applies to groups $G$ which may not be finitely generated themselves. For instance, if $Q$ is the additive group of rational numbers, which is an extension of the integers (a finitely generated, infinite normal subgroup) with infinite quotient group, we see that $E(Q) \approx Z_2$. We could therefore derive a result such as this: Let $A$ be any subset of $Q$ such that, for all $n$, $A$ differs from

$$A + \frac{1}{n!}$$

by only some finite set. Then, either $A$ is itself finite, or $Q - A$ is finite.

This is also an instance of a group with one end being a direct limit of groups with two ends.

Other peculiar cases of infinitely generated groups can be considered here. In the additive group $G = Z_{n^\infty}$ of those rationals mod 1 whose denominators can be expressed as powers of $n$, for $n > 2$ we can describe a nontrivial element of $E(G)$ by the element $A \in \mathcal{Q}(G)$ consisting of all elements of $G$ that can be written in the form

$$\frac{a}{n^k} \quad \text{where} \quad a \equiv 1 \pmod{n}.$$

It is easily proved that the difference in the sets

$$A \quad \text{and} \quad A + \frac{1}{n^s}$$

is finite and, since

$$\left\{\frac{1}{n^s}\right\}$$

generates $G$, this shows $A \in \mathcal{Q}(G)$; and, clearly, both $A$ and $A^*$ are infinite.

From this example we can generalize to show that $Z_{n\infty}$ has infinitely many ends. This kind of group gives counterexamples to show that "finitely generated" cannot be dispensed with in theorems 4.A.6.3, 4.A.6.4, and 5.A.10.

Proof of 4.A.6.4 G finitely generated, $A \in \mathcal{E}(G)$, nontrivial, such that there are infinitely many $g \in G$ with $Ag = A$, implies G has two ends.

A is represented by an element, also called A, in $\mathcal{Q}(G)$ such that neither A nor $A^*$ is finite, and, for infinitely many g, $Ag + A$ is finite. We consider the graph $\Gamma$ of G relative to a finite set of generators; in this we have $\delta A$ finite, since $A \in \mathcal{Q}$. There is therefore a finite connected subgraph $\Delta$ of $\Gamma$, containing all the edges of $\delta A$.

Let $\Gamma_A$ be the graph consisting of $\Delta$ and all the vertices in A and all edges whose vertices are in A or $\Delta$. Then $\Gamma_A$ is a connected graph; otherwise there would be a proper subset of A with empty coboundary.

Since G acts freely on $\Gamma$, there are only finitely many $g \in G$ such that $\Delta \cap \Delta g \neq \emptyset$. Hence there is $g \in G$ with $A + Ag$ finite and $\Delta \cap \Delta g = \emptyset$; for this g, not both $A - Ag$ and $Ag - A$ can be nonempty (Figure 4.1).

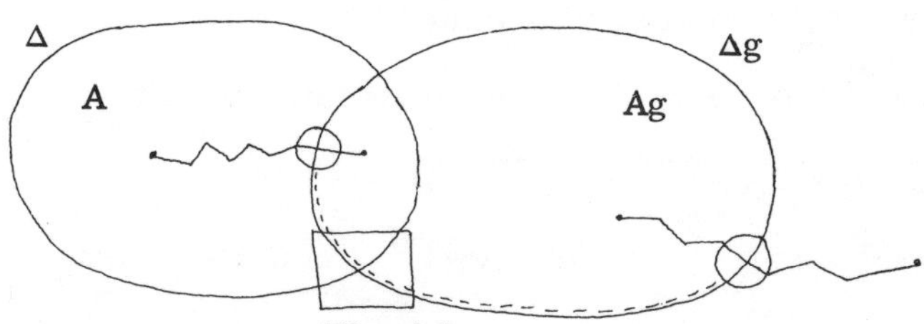

Fig. 4.1

The argument for this is based on geometric reasoning; we must have both $A \cap Ag$ and $A^* \cap A^*g$ infinite and hence nonempty. If $A - Ag \neq \emptyset$, then a path in $\Gamma_A$ joining points in $A - Ag$ and $A \cap Ag$ exists and hence must intersect $\delta A_g \subset \Delta g$; the place where the intersection occurs happens well within A. A similar argument in $A^*$ shows, if $Ag - A \neq \emptyset$, that some part of $\Delta g$ lies well within $A^*$. A path connecting these two, within $\Delta g$, since $\Delta g$ is connected, will now intersect $\delta A \subset \Delta$; and so $\Delta \cap \Delta g$ would not be $\emptyset$.

Therefore, either $A \subset Ag$ or $Ag \subset A$. We can, by looking at $g^{-1}$, if necessary, assume $A \subset Ag$, and we have the further fact, since $\Delta \cap \Delta g = \emptyset$, that $\delta A \cap \delta Ag = \emptyset$. Let $B = Ag - A$; this is finite, and on looking at

$$C = \bigcup_{n=-\infty}^{+\infty} Bg^n$$

we can see easily that $\delta C = \emptyset$. It follows that $C = G$ then.

Suppose now $D$ is any element of $Q(G)$. Since $\delta D$ is finite, there is $k$ so that all vertices of some finite connected graph $\Delta'$ containing $\delta D$ lie in

$$C_k = \bigcup_{n=-k}^{+k} Bg^n.$$

An argument similar to the one involving $A$ and $Ag$ shows that there are four possibilities:

$D \subset C_k$ so that $D$ is finite.

$D^* \subset C_k$ so that $D^*$ is finite.

$Ag^{-k} \subset D$ and $D - Ag^{-k}$ is finite so that

$D + A = D + Ag^{-k} + (Ag^{-k} + A)$ is finite.

$A^*g^{k+1} \subset D$ and $D - A^*g^{k+1}$ is finite, so that

$D + A^*$ is finite.

Thus an element of $Q(G)$ can represent only one of four elements of $E(G)$, and $G$ has then exactly two ends.

Comment on 4.A.6.4 If we look upon $E(G)$ as a set on which $G$ acts and suppose $G$ has more than two ends and that $G$ is finitely generated, then 4.A.6.4 says the isotropy group $I$ of any nontrivial element of $E(G)$ is finite. Therefore the orbit of any nontrivial element which is in 1-to-1 correspondence with the cosets G/I is infinite. This implies $E(G)$ is infinite and that G acts in a highly nontrivial way on $E(G)$. Thus we obtain the theorem that a finitely generated group with more than two ends has infinitely many ends.

Proof of 4.A.6.5 To say that G is finitely generated, having two ends, implies there is a finite normal subgroup $N \lhd G$, such that $G/N \approx Z$ or $\approx Z_2 * Z_2$.

Let A be an element of $Q(G)$ such that A and A* are both infinite. For all $g \in G$, either $Ag + A$ is finite or $Ag + A^*$ is finite. The set H of those $g \in G$ with $Ag + A$ finite is a subgroup of index 1 or 2 in G.

We define now a function $\phi : H \to Z$ as follows. Let $f : G \to Z$ have $f(g) = 1$ for $g \in A$ and $f(g) = 0$ for $g \in A^*$. Let, for $g \in H$, $f_g(a) = 1$ if $a \in Ag$, and $= 0$ if $a \in A^*g$. Define

$$\phi(h) = \int(f_h - f);$$

i.e., $\phi(h)$ = the number of elements of $Ah - A$ minus the number of elements of $A - Ah$. We see $\phi$ is a homomorphism, by realizing that $f_{hg} - f_g$ is the translation of $f_h - f$ by g and so has the same integral, so that

$$\phi(hg) = \int(f_{hg} - f) = \int(f_{hg} - f_g) + \int(f_g - f) = \phi(h) + \phi(g).$$

The kernel N of $\phi$ is seen, by considerations in the proof of 4.A.6.4, to be finite, and so $\phi$ maps onto an infinite cyclic subgroup of Z.

If H has index 2 in G, then it is possible to extend $\phi$ into the finite dihedral group by taking account of the possibility that $Ag \equiv A^*$ modulo finite sets.

4.A.8 Computations and the Proof of 4.A.6.6

To make computations of ends, it is easiest to work with groups with a bipolar structure.

Let G be such a group and $\{F, S, EE, EE^*, E^*E, E^*E^*\}$ be a bipolar structure on it. By the nontriviality axiom, $EE^* \neq \emptyset$; it follows from axiom 5 that $E^*E \neq \emptyset$, and from axiom 6 that $EE^*$ and $E^*E$ are both infinite.

Let $A = EE \cup E^*E$. It follows that both A and $A^* = G - A$ are infinite. Let g be any irreducible element of the bipolar structure. If $g \in F \cup S$, then by axioms 3 and 4, $gA = A$. Otherwise, by 3.B.2.2 and 3.B.2.3,

$$gA \subset A \cup F \cup S,$$

and, since $g^{-1}$ is also irreducible,

$$g^{-1}A \subset A \cup F \cup S,$$

so that $A \subset gA \cup g(F \cup S)$. Thus

$$gA - A \subset F \cup S \quad \text{and} \quad A - gA \subset g(F \cup S).$$

Hence $\nabla_g(A) = A + gA \subset F \cup S \cup g(F \cup S)$ is finite for all irreducible elements g. Since G is generated by its irreducible elements, therefore by 4.A.4, $A \in \mathcal{Q}(G)$; and A represents a nontrivial element of $\mathcal{E}(G)$ since A and $A^*$ are both infinite. Thus:

4.A.8.1 If G has a bipolar structure, then G has more than one end.

Let $G_2$ be the subgroup of G consisting of $F \cup \{$irreducible elements of $E^*E^*\}$. If we suppose that F is of index $\geq 3$ in $G_2$, we can find $a, b \in G_2 - F$ and $ab^{-1} \notin F$. Then consider $A = EE \cup E^*E$ as above; the sets A, Aa, Ab are then all disjoint elements of $\mathcal{Q}(G)$ and thus represent three distinct nontrivial elements of $\mathcal{E}(G)$. We combine this with .6.4 to see:

4.A.8.2 If G has a bipolar structure and F is of index $\geq 3$ in $G_2$, then G has infinitely many ends.

A rather similar computation can be made when $S = \emptyset$; $EE^*$ contains an irreducible element x, and $G_2 - F$ contains some element a. In this case, xa is an irreducible element of $EE^*$; $Ax^{-1}$, $Aa^{-1}x^{-1}$, and $A^*$ are then disjoint. Therefore, as in the earlier case, G has infinitely many ends:

4.A.8.3 If $G$ has a bipolar structure, $S = \emptyset$, $F$ of index $\geq 2$ in $G_2$, and $EE^*$ contains some irreducible element, then $G$ has infinitely many ends.

We now describe some particular instances of bipolar structures.

(i) Let $G = G_1 *_F G_2$ where $F$ is a finite group properly contained in each of $G_1$ and $G_2$. Thus $G$ is the universal group of the pregroup $G_1 \cup G_2$; considering words in that pregroup, if $X$ and $Y$ stand for $E$ or $E^*$, let $XY$ consist of all those elements of $G$ represented by reduced words $(x_1, \ldots, x_n)$ in $G_1 \cup G_2$, where

if $X = E$, $x_1 \in G_1 - F$
if $X = E^*$, $x_1 \in G_2 - F$
if $Y = E$, $x_n \in G_1 - F$
if $Y = E^*$, $x_n \in G_2 - F$.

It is easily verified that the resultant structure, with $S = \emptyset$, is a bipolar structure on $G$.

(ii) Let $G = G_1 \, {}_F\mho\phi$ where $F$ is a finite subgroup of $G_1$. Then $G$ is the universal group of the pregroup $P$ consisting of $G_1$, $xG_1$, $G_1x^{-1}$, $xG_1x^{-1}$, where $F$ is identified with $x\phi(F)x^{-1}$. We construct a bipolar structure, making $XY$ consist of elements represented by reduced words $(u_1, \ldots, u_n)$ in $P$ such that

if $X = E$, $u_1 \in (xG_1x^{-1} - F) \cup xG_1$
if $X = E^*$, $u_1 \in (G_1 - F) \cup G_1x^{-1}$
if $Y = E$, $u_n \in (xG_1x^{-1} - F) \cup G_1x^{-1}$
if $Y = E^*$, $u_n \in (G_1 - F) \cup xG_1$.

This too produces a bipolar structure with $S = \emptyset$.

Now, finitely generated groups with two ends have been characterized in 4.A.6.5, and it results that each such group $G$ can be written as either of two cases:

(i) There is the case $G = G_1 *_F G_2$ where $F$ is of index 2 in both $G_1$ and $G_2$. This is the case when $G$ has a finite normal subgroup $F$ with quotient group $Z_2 * Z_2$. This latter group has a subgroup of index 2 which is infinite cyclic; the infinite cyclic group has two ends clearly, and so by 4.A.6.1 and 4.A.6.2, any group of this form definitely has 2 ends.

(ii) Or there is the case where $G = F\ {}_F\circlearrowleft\phi$, where F is finite and $\phi$ is some automorphism of F. This is the case when G has a finite normal subgroup F with infinite cyclic quotient group. Every such group has two ends.

We now know exactly how many ends G has if it can be written as $G = G_1 *_F G_2$ or $G = G_1\ {}_F\circlearrowleft\phi$. If it falls into one of the immediately preceding two cases, then G has two ends; while if these cases are not so, then the corresponding bipolar structure satisfies the hypothesis of A.4.B.2 or A.4.8.3, and so G has infinitely many ends.

In particular, this completes the proof of 4.A. 6.6.

## 4.B Results in Graph Theory

A graph $\Gamma$ consists of two sets, $\Gamma_0$ and $\Gamma_1$ of "vertices" and "edges" and two functions $v_1, v_2 : \Gamma_1 \to \Gamma_0$, "first" and "second" vertex. The graph $\Gamma$ is said to be locally finite if $v_1$ and $v_2$ have the property that the inverse image of each element of $\Gamma_0$ is finite.

A typical example is the graph of a group G with respect to a set $T \subset G$. Here $\Gamma_0 = G$, $\Gamma_1 = T \times G$, and $v_1(t,g) = g$, $v_2(t,g) = tg$. The graph $\Gamma$ is locally finite if T is finite, and it is connected (in the obvious sense) if T generates G.

The Boolean algebra of all subsets of $\Gamma_0$ is called $A(\Gamma)$; AB denotes intersection. The group of all subsets of $\Gamma_1$ under symmetric difference is called $B(\Gamma)$; if necessary we write $X \cap Y$ for the intersection of two subsets of $\Gamma_1$ since we do not conceive of $B(\Gamma)$ in the form of a Boolean algebra.

If $A \in A(\Gamma)$, $X \in B(\Gamma)$, we define

$$AX = \{e \in X \mid v_1(e) \in A\}$$

$$XA = \{e \in X \mid v_2(e) \in A\}.$$

These are elements of $B(\Gamma)$; for $A, B \in A(\Gamma)$ and $X, Y \in B(\Gamma)$ we have associative and distributive rules for $ABX$, $AXB$, $XAB$, $(A + B)X$, $A(X + Y)$, $X(A + B)$, $(X + Y)A$.

We denote by $1 \in A(\Gamma)$, the set $\Gamma_0$ itself, for which we have rules $1A = A1 = A$, $1X = X1 = X$; and denote $A^* = 1 + A$ = complement of $A$.

For $A \in A(\Gamma)$, we define $\delta A \in B(\Gamma)$, thus:

$$\delta A = A\Gamma_1 + \Gamma_1 A.$$

We have the rules:

$$\delta(A + B) = \delta A + \delta B$$
$$\delta(AB) = (\delta A)B + A(\delta B)$$
$$\delta(A^*) = \delta A.$$

An element $A \in A(\Gamma)$ is said to be connected if, whenever $A = B + C$, with $BC = 0$ and $B \neq 0 \neq C$, then $(\delta B) \cap (\delta C) \neq 0$. This is the same as requiring that there exist a connected subgraph of $\Gamma$ having vertices $A$.

The entire graph $\Gamma$ is connected if and only if 1 is connected, if and only if $\delta A = 0$ implies $A = 0$ or $A = 1$.

4.B.1 If $A, B \in A(\Gamma)$, with $0 \neq A \subset B$ and $\delta A \subset \delta B$, and $B$ is connected, then $A = B$.

Proof: $B = A + (A + B)$, and, since $A \subset B$, we have $A(A + B) = 0$. Thus, if $A + B \neq 0$, it must happen that $\delta A \cap \delta(A + B) \neq 0$, but this contradicts $\delta A \subset \delta B$.

Let $Q(\Gamma)$ denote the set of those $A \in A(\Gamma)$ such that $\delta A$ is finite. Then $Q(\Gamma)$ is a subalgebra of $A(\Gamma)$. Let $|\delta A|$ denote the number of elements of $\delta A$. Let $M$ be the subgroup of $B(\Gamma)$ consisting of all $\delta A$ for $A$ finite. Recalling the definition of the ends of a graph in 4.A, we see that, if $\Gamma$ has more than one end, then $M$ is definitely smaller than $\delta(Q(\Gamma))$. The following discussion could perhaps be extended by being less specific about $M$, but we have not done this in detail and so do not include it here.

An element $A \in Q(\Gamma)$ such that $\delta A \notin M$, will be called nontrivial. Let $k$ be the minimum of all $|\delta A|$ for $A$ nontrivial, and call $k$ the width of $\Gamma$.

If $\Gamma$ is connected and has more than one end, then its width is a well-determined positive integer. An element $A \in \mathcal{A}(\Gamma)$ which is nontrivial and whose $|\delta A| = k$, the width of $\Gamma$, will be called narrow. We went to derive some lattice-theoretic properties of the set of narrow elements of $\mathcal{Q}(\Gamma)$.

4.B.2 If A is narrow and $\Gamma$ is connected, then A is connected.

Proof: If not, A could be written as a nontrivial disjoint sum B + C whose boundaries were disjoint. Since $\delta A = \delta B + \delta C \notin M$, either $\delta B$ or $\delta C$ is not in M, say $\delta B$; since $0 \neq C \neq 1$ and $\Gamma$ is connected, $\delta C \neq 0$ and so $|\delta B| < |\delta A|$; thus A would not be narrow.

4.B.3 (Descending Chain Condition) Let $A_1 \supset A_1 \supset \ldots$ be a descending sequence of narrow elements of $\mathcal{Q}(\Gamma)$, where $\Gamma$ is connected. Suppose

$$B = \bigcap_{n=1}^{\infty} A_n \neq 0.$$

Then $B = A_n$ for some n.

Proof: Let k be the width of $\Gamma$. If $e \in \delta B$, then one vertex of e belongs to all $A_n$ and the other is outside of all but finitely many $A_n$. Thus, there is $n_e$, such that for all $n \geq n_e$, $e \in \delta A_n$. Hence there are no more than k elements in $\delta B$ since, if there were more, then almost all $A_n$ would have at least k + 1 elements in $\delta A_n$. It follows that $\delta B$ is contained in some $\delta A_n$. By 4.B.2, $A_n$ is connected, and so, by 4.B.1, $B = A_n$.

4.B.4 Corollary There exists a narrow element $A \in \mathcal{Q}(\Gamma)$ which is minimal with respect to containing $v \in \Gamma_0$. That is, if $v \in B \subset A$ and B is narrow, then B = A.

Otherwise there would (by the axiom of choice) exist a descending sequence of narrow sets, all containing v.

4.B.5 Theorem If A is narrow, minimal with respect to containing v, and B is any narrow element, then at least one of the following

$$X = AB,\ AB^*,\ A^*B,\ A^*B^*$$

has $\delta X \in M$.

Proof: Suppose none has $\delta X \in M$. Let k be the width of $\Gamma$; write:

$$\delta(AB) = A(\delta B) + (\delta A)B$$
$$\delta(AB^*) = A(\delta B) + (\delta A)B^*$$
$$\delta(A^*B) = A^*(\delta B) + (\delta A)B$$
$$\delta(A^*B^*) = A^*(\delta B) + (\delta A)B^*$$

using the fact that $\delta A = \delta A^*$.

We deduce from this table that

$$|\delta(AB)| + |\delta(AB^*)| + |\delta(A^*B)| + |\delta(A^*B^*)| \leq 2|\delta A| + 2|\delta B| = 4k.$$

We cannot have any $|\delta X| < k$ since then, having $\delta X \notin M$, the width of $\Gamma$ would be less than k. Thus all have $|\delta X| \geq k$ and hence $= k$. Thus all the sets

$$AB, \ AB^*, \ A^*B, \ A^*B^*$$

are M-narrow. However, in that case, one of AB or $AB^*$ contains v and is properly smaller than A, in contradiction to the minimality of A.

We shall summarize this discussion in one self-contained statement.

4.B.6 If $\Gamma$ is a locally finite, connected graph with more than one end, let

$$k = \min \{ |\delta A| \mid A \in Q(\Gamma), \ A \text{ and } A^* \text{ both infinite} \};$$

call A narrow if A and $A^*$ are both infinite and $|\delta A| = k$. Then there exists a narrow A such that, for all narrow B, one of

$$AB, \ AB^*, \ A^*B, \ A^*B^*$$

is finite.

Proof: We take A to be a narrow element which is minimal with respect to containing some particular vertex. By 4.B.5, one of $\delta(AB)$, etc., say $\delta(AB)$, is $\delta(F)$ for some finite set F; hence, $\Gamma$ being connected, $AB + F = 0$ or 1; it is impossible to have $AB + F = 1$ since $A^*$ is infinite and F is only finite; and so $AB + F = 0$ or $AB = F$.

4.B.7 Let A and B be narrow elements of $Q(\Gamma)$ such that AB and $A^*B^*$ are infinite. Then AB and $A \cup B = (A^*B^*)^*$ are narrow.

Proof: Let k be the width of $\Gamma$. We note that AB and $(AB)^* \supset A^*B^*$ are both infinite. Hence $|\delta(AB)| \geq k$; similarly $|\delta(A^*B^*)| \geq k$. Writing

$$\delta(AB) = (\delta A)B + A(\delta B)$$
$$\delta(A^*B^*) = (\delta A^*)B^* + A^*(\delta B^*) = (\delta A)B^* + A^*(\delta B)$$

we see that

$$|\delta(AB)| + |\delta(A^*B^*)| \leq |\delta A| + |\delta B| = 2k.$$

Hence $|\delta(AB)| = |\delta(A^*B^*)| = k$ and both AB and $A^*B^*$ are narrow, and, since $\delta(A \cup B) = \delta(A^*B^*)$, it follows that $A \cup B$ is narrow.

We recall that $E(\Gamma)$ is the quotient algebra of $Q(\Gamma)$ by the ideal of finite subsets of $\Gamma_0$.

4.B.8 Let $0 \neq \alpha \subset \beta \neq 1$, where $\alpha, \beta \in E(\Gamma)$. Let $L(\alpha,\beta) = \{\gamma \in E(\Gamma) \mid \alpha \subset \gamma \subset \beta$ and $\gamma$ is representable by a narrow element C of $Q(\Gamma)\}$.

Then $L(\alpha,\beta)$ is closed under intersection and union, and, being a sublattice of a Boolean algebra, is a distributive lattice.

The proof is based on picking representatives and applying 4.B.7.

4.B.9 The set $L(\alpha,\beta)$ satisfies the descending chain condition. I.e., if $\beta \supset \gamma_1 \supset \gamma_2 \supset \dots \supset \gamma_n \supset \dots \supset \alpha$, where all $\gamma_i$ are representable by narrow elements $C \in Q(\Gamma)$, there exists N, such that for $n \geq N$, $\gamma_n = \gamma_N$.

Proof: We can suppose for all n, $\gamma_n \neq \alpha$, otherwise the conclusion is clear. Represent $\gamma_i$ by $C_i \in Q(\Gamma)$ with $C_i$ narrow. Then $\gamma_i$ is also representable by $D_i = C_1 \cap \dots \cap C_i$, which is narrow by 4.B.7, and we have $D_1 \supset D_2 \supset \dots \supset D_n \supset \dots$.

Represent $\alpha$ by $A \in Q(\Gamma)$. Let S denote the set of vertices of edges in $\delta A$; then S is a finite set. Now, $D_i$, being narrow, is, by 4.B.2, connected; furthermore, since $\gamma_i\alpha^* \neq 0$ (since $\gamma_i \neq \alpha$) and $\gamma_i\alpha = \alpha \neq 0$, we have $D_iA^* \neq 0 \neq D_iA$. It follows from connectedness of $D_i$ that the two sets

$$\delta(D_iA^*) = (\delta D_i)A^* + D_i(\delta A) \quad \text{and} \quad \delta(D_iA) = (\delta D_i)A + D_i(\delta A)$$

must intersect, and hence $D_i(\delta A) \neq 0$ and so $D_i \cap S \neq \emptyset$. Therefore,

$$\bigcap_{n=1}^{\infty} D_n \supset \bigcap_{n=1}^{\infty} (D_n \cap S) \neq \emptyset$$

since an intersection of a decreasing sequence of nonempty finite sets is nonempty. By 4.B.3, then,

$$\bigcap_{n=1}^{\infty} D_n = D_N$$

for some N, and so for $n \geq N$, $D_n = D_N$, and so $\gamma_n = \gamma_N$.

4.B.10 Dually, $L(\alpha,\beta)$ satisfies the ascending chain condition.

Proof: This is equivalent to the descending chain condition on $L(\beta^*,\alpha^*)$.

4.B.11 (Boundedness of chains in $L(\alpha,\beta)$). Given $0 \neq \alpha \subset \beta \neq 1$ in $E(\Gamma)$, there is $N = N(\alpha,\beta)$ such that whenever

$$\alpha \subset \gamma_1 \subset \gamma_2 \subset \ldots \subset \gamma_n \subset \beta$$

$\gamma_i \neq \gamma_{i+1}$ for $i = 1, \ldots, n-1$, and $\gamma_i$ representable by narrow elements of $Q(\Gamma)$ for $i = 1, \ldots, n$, then $n \leq N$.

Proof: Because of the ascending and descending chain conditions on $L(\alpha,\beta)$, there is a chain in $L(\alpha,\beta)$ which cannot be further enlarged:

$$\alpha \subset \delta_1 \subset \delta_2 \subset \ldots \subset \delta_N \subset \beta$$

such that if $\delta_i \subset \varepsilon \subset \delta_{i+1}$, $\varepsilon \in L(\alpha,\beta)$, then $\delta_i = \varepsilon$ or $\varepsilon = \delta_{i+1}$.

Now, if there were a longer chain

$$\alpha \subset \gamma_1 \subset \ldots \subset \gamma_n \subset \beta \quad \text{with} \quad n > N,$$

we could, by the Jordan-Hölder-Schreier-Zassenhaus-Artin theorem applied to $L(\alpha,\beta)$, refine the $\delta$-chain to contain at least n−1 nontrivial inclusions. This would contradict the maximality of the $\delta$-chain.

# 5 CONCLUSIONS

## 5.A The Structure of Groups with Infinitely Many Ends

Let G be a finitely generated group with more than one end. We have seen exactly what kinds of groups have two ends, in 4.A.6.5; namely, a group which has a homomorphism with finite kernel onto Z or $Z_2 * Z_2$. If G has more than two ends, then, by 4.A.6.4, it has infinitely many ends.

Let T be a finite set generating G and $\Gamma$ the corresponding graph; we note that G acts on $\Gamma$ on the right and hence acts on $\mathcal{Q}(\Gamma)$ and $\mathcal{E}(\Gamma)$ on the right.

There is, by 4.B.6, an element $A \in \mathcal{Q}(\Gamma)$ which is narrow and which is minimal among narrow sets that contain some particular $v \in G = \Gamma_0$. Given $g \in G$, we note that $|\delta A| = |\delta Ag|$ so that Ag is narrow, and thus, by 4.B.6, at least one of the four sets

$$Ag \cap A,\ Ag \cap A^*,\ A^*g \cap A,\ A^*g \cap A^*$$

is finite. Since A, A*, Ag, A*g are all infinite, it is impossible that four of the six possible pairs of these sets consist of finite sets. The question of which ones of these sets are finite thus partitions G into six sets. We shall have to deal with the notation as follows:

Let E and E* be operators on $\mathcal{A}$ as follows: E(A) = A, E*(A) = A*; let EE, EE*, E*E, E*E* be the four symbols made by concatenating two of $\{E, E^*\}$; let X, Y, Z, etc., be variables standing for E or E* with the convention that, if X stands for E or E*, then X* stands for E* or E, respectively.

Define F, S, EE, etc., as subsets of G, as follows:

$$F = \{g \in G \mid \text{both } Ag \cap A^* \text{ and } A^*g \cap A \text{ are finite}\}$$
$$= \{g \in G \mid A = Ag \text{ in } E(G)\}$$
$$S = \{g \in G \mid \text{both } Ag \cap A \text{ and } A^*g \cap A^* \text{ are finite}\}$$
$$= \{g \in G \mid Ag = A^* \text{ in } E(G)\}.$$
$$H = F \cup S$$
$$XY = \{g \in G - H \mid X(Ag) \cap Y(A) \text{ is finite}\}$$
$$= \{g \in G \mid X(Ag) \subsetneq Y^*(A) \text{ in } E(G)\}$$
$$= \{g \in G \mid Y(A) \subsetneq X^*(Ag) \text{ in } E(G)\}.$$

These last three equivalent conditions define four subsets EE, EE*, E*E, E*E* of A.

We hereby explicitly assume that we have a group G with infinitely many ends.

5.A.1 F is a finite subgroup of G.

This follows from 4.A.6.4 since F is the isotropy group of the element A in $E(G)$.

5.A.2 $H = F \cup S$ is a subgroup of G in which F has index 1 or 2.

This is clear.

5.A.3 If $g \in XY$, then $g^{-1} \in YX$.

For, $X(Ag) \subsetneq Y^*(A)$ in $E(G)$. Multiply on the right by $g^{-1}$, to get $X(A) \subsetneq Y^*(Ag^{-1})$, which implies

$$g^{-1} \in YX.$$

5.A.4 If $g \in XY$ and $f \in F$, then $gf \in XY$.

For, in $E(G)$, $X(Ag) \subsetneq Y^*(A)$, and $Y^*(Af) = Y^*(A)$. So, by multiplying the first inequality by f:

$$X(Agf) \subsetneq Y^*(Af) = Y^*(A)$$

and hence $gf \in XY$.

5.A.5 If $g \in XY$ and $s \in S$, then $gs \in XY^*$.

For, in $E(G)$, $Y^*(As) = Y^*(A^*) = Y(A)$, so that

$$X(Ag) \subsetneq Y^*(A) \quad \text{implies} \quad X(Ags) \subsetneq Y^*(As) = Y(A)$$

and $gs \in XY^*$.

5.A.6 If $g \in XY$ and $p \in Y^*Z$, then $gp \in XZ$.

For, we have, in $E(G)$,

$$X(Ag) \subsetneq Y^*(A) \quad \text{so that} \quad X(Agp) \subsetneq Y^*(Ap)$$

And also

$$Y^*(Ap) \subsetneq Z^*(A).$$

Hence,

$$X(Agp) \subsetneq Z^*(A) \quad \text{or} \quad gp \in XZ.$$

We have at this point proved all the axioms of a bipolar structure, except for the boundedness assertion and the nontriviality axiom, axioms 7 and 8.

Let $X_1, \ldots, X_{n+1}$ be symbols, and let $g_i \in X_i X_{i+1}^*$, so that for all $i$,

$$X_i(Ag_i) \subsetneq X_{i+1}(A)$$

or

$$X_1(Ag_1g_2 \ldots g_n) \subsetneq X_2(Ag_2 \ldots g_n) \subsetneq \ldots \subsetneq X_n(Ag_n) \subsetneq X_{n+1}(A).$$

We recognize now the fact that there is an upper bound on $n$, given $p = g_1g_2 \ldots g_n$, is a consequence of 4.B.11, that chains of narrow elements between two narrow elements have a limited size. Thus:

5.A.7 For given $p \in G$, there is $N(p)$ such that, if $p = g_1g_2 \ldots g_n$ and $g_i \in X_i X_{i+1}^*$, then $n \leq N(p)$.

5.A.8 $EE^*$ is not empty.

<u>Proof</u>: Let $\Delta$ be a finite connected subgraph of $\Gamma$ containing all the edges of $\delta A$. Let $L$ be the set of vertices of $\Delta$. It is easy to see, be-

cause of the fact that G acts transitively on vertices of $\Gamma$, and $\Delta$ has finite size, that there exist $x, y \in G - H$, such that

$$\Delta x \cap \Delta = \emptyset = \Delta y \cap \Delta \quad \text{and} \quad Lx \subset A,\ Ly \subset A^* .$$

We cannot have, if $\Delta x \cap \Delta = \emptyset$, all four of the sets $(Ax) \cap A$, $(Ax) \cap A^*$, etc., nonempty. (To prove this is to argue along the lines of the proof of 4.A.6.4.) By construction, since L contains points both in A and in $A^*$, we must have $(Ax) \cap A$ and $(A^*x) \cap A$ nonempty and also $(Ay) \cap A^*$ and $(A^*y) \cap A^*$ nonempty. It follows that either $(Ax) \cap A^*$ or $(A^*x) \cap A^*$ is empty, and, since $x \notin H$, we have $x \in EE^*$ or $x \in E^*E^*$; similarly, $y \in EE$ or $y \in E^*E$.

Therefore, one of the elements $x, y^{-1}$, yx belongs to $EE^*$.

Then, we have proved that, given a minimal narrow $A \in \mathcal{Q}(\Gamma)$, where $\Gamma$ is the graph of the group G with respect to a finite set of generators, and, assuming that G has more than two ends, we can define a bipolar structure on G.

5.A.9 If G is a finitely generated group with infinitely many ends, then there is a bipolar structure on

$$G : \{F, S, EE, EE^*, E^*E, E^*E^*\}.$$

Thus G has one of the following two types of structure:

(1) $G = G_1 *_F G_2$, a free product with finite amalgamated subgroup F, properly contained in both factors, and of index $> 2$ in at least one factor.

(2) $G = G_1\, {}_F\mathfrak{I}\phi$, where F is a finite subgroup, properly embedded in $G_1$.

<u>Proof</u>: These are just the possibilities for a bipolar structure listed in 3.B.5, taking into account 4.A.8 to exclude the cases where G has two ends. Adding to this the result stated as 4.A.6.6, we have an if and only if statement:

5.A.10 A finitely generated group has infinitely many ends, if and only if it can be decomposed according to case (1) or (2) of 5.A.9.

## 5.B Group-Theoretic Consequences

A torsion-free group is defined as one in which the only element of finite order is the identity.

5.B.1 A finitely generated, torsion-free group has two ends if and only if it is infinite cyclic. It has infinitely many ends if and only if it can be written nontrivially as a free product.

This follows from 4.A.6.5 and 5.A.10.

5.B.2 If $P$ is a property of groups which is inherited by nontrivial free factors and if when a group has property $P$ then is has more than one end, then any finitely generated torsion-free group with property $P$ is a free group.

Proof: Let G have property $P$. Then G has more than one end; if G has two ends, by 5.B.1 it is free (cyclic). Otherwise, by 5.B.1, it is a free product $G_1 * G_2$, and both $G_1$ and $G_2$ have property $P$ and (by Grushko's theorem) can be generated by fewer generators than G can; by induction on the number of generators, $G_1$ and $G_2$ are free, and hence G is free.

5.B.3 Serre Conjecture If G is a finitely generated torsion-free group with a free subgroup F of finite index, then G is a free group.

Proof: Any finitely generated subgroup H of G contains the free group $F \cap H$ as a subgroup of finite index. A nontrivial free group has more than one end, and so, by 4.A.6.2, any supergroup of finite index containing it has more than one end. Thus by 5.B.2, G is free.

5.B.4 Eilenberg-Ganea Conjecture If G is a finitely generated group of cohomological dimension 1, then G is free.

Proof: The property $P$ of having cohomological dimension 1 is inherited by nontrivial subgroups, and implies that group is torsion free. That it implies the group has more than one end is a simple consequence [see 27] of the duality theory for finitely generated projective modules over the group ring of G, together with homological interpretation of ends.

## 5.C The Sphere Theorem

Finally, we propose to sketch a proof of the sphere theorem. This is presented not so much as an alternative to the proof of Papakyriakopoulos, Whitehead, and Epstein as it is a demonstration of the interweaving of topological and group-theoretic facts. Thus, in the arguments of Papakyriakopoulos there appear ideas about covering spaces and the disentangling of singularities which suggest a whole combinatorial investigation into group theory, such as this paper has begun. Having started this investigation, we have now attained a point where the sphere theorem is easier for the intuition to grasp.

We shall not quite prove the most general form of the sphere theorem here, instead proving a fairly general form from which the theorem in 1.B.1 can be derived without great difficulty. We, of course, assume the Loop Theorem, Dehn's Lemma, and the other consequences of Chapter 2.

5.C.1 Let $M$ be a compact 3-manifold such that $\pi_2(M) \neq 0$. Then there is in $M$ a two-sided $\Sigma$ which is a 2-sphere or projective plane, which carries a nontrivial element of $\pi_2(M)$.

Proof: We first simplify the picture by making modifications along the boundary, which allow us to conclude that the fundamental group of $M$ will, in this simple form, have more than one end. Since we have a characterization of such groups, and therefore a topological picture of such groups, we invent a situation to which we can apply 2.B.2, and, after some topological modifications of this, we can then see the conclusion of the theorem.

If $\partial M$ contains a 2-sphere or projective plane, then that boundary component itself determines a nontrivial element of $\pi_2(M)$ and, pushed slightly into $M$, can be taken as $\Sigma$. (It is easy to see that, if $\partial M$ contains a 2-sphere which is contractible in $M$, then $M$ is itself contractible.)

If, for some component $T$ of $\partial M$, the map $\pi_1(T) \to \pi_1(M)$ is not injective, we can apply the Loop Theorem and Dehn's Lemma to write $M$ as $M'$ plus a handle $D^2 \times I$; since $M$ has the homotopy type of the wedge of $M'$ and a circle, an argument on the homology of the universal covering spaces shows $\pi_2(M') \neq 0$. The operation $M \to M'$ performs a reduction on $\partial T$.

Thus either the theorem is true or there is a submanifold $M''$ of $M$, with $\pi_2(M'') \to \pi_2(M)$ injective and $\pi_2(M'') \neq \{0\}$ and $\partial M''$ containing no 2-spheres or projective planes and $\pi_1(\partial M'') \to \pi_1(M'')$ injective.

In this case, if $\widetilde{M}''$ is the universal covering space of $M''$, $\partial\widetilde{M}''$ consists of copies of the universal covers of $\partial M''$ and so is acyclic in dimensions $\geq 1$. So $\pi_2(M'') \approx H_2(\widetilde{M}'') \approx H_2(\widetilde{M}'', \partial\widetilde{M}'') \approx H_f^1(\widetilde{M}'')$. If $C_f$ and $C$ denote the cochain complexes of finite and ordinary cochains respectively, and $C_e$ their quotient, then we have an exact cohomology sequence, which implies that $H_e^0(\widetilde{M}'')$ has rank $\geq 2$; this is on the basis of the facts that $H^1(\widetilde{M}'') = 0$ and $H_f^0(\widetilde{M}'') = 0$; these follow from the facts that $\widetilde{M}''$ is a universal covering space and $\pi_1(M'')$ is infinite, otherwise $\widetilde{M}''$ would either be a homotopy 3-sphere or all components of $\partial M''$ would have finite fundamental groups. For these reasons we conclude, cf. [12], [21], that $\widetilde{M}''$ and hence $\pi_1(M'')$ has at least two ends. On the basis of 3.B.6, 4.A.6.5, 4.A.6.6, and the theorem 5.A.10, if $G = \pi_1(M'')$, we can construct an aspherical space having fundamental group $G$ as follows:

Corresponding to case (1) of 5.A.9, $G = G_1 *_F G_2$, we let $K_1$ and $K_2$ be aspherical spaces with fundamental groups $G_1$, $G_2$, containing a copy of $K_F$ an aspherical space with fundamental group $F$. We let $X =$ $K_1 \cup \{K_F \times [-1, +1]\} \cup K_2$, identifying $K_F \times (-1)$ with $K_F$ in $K_1$ and $K_F \times (+1)$ with $K_F$ in $K_2$. Define $A$ as $K_F \times 0$.

Corresponding to case (2) of 5.A.9, $G = G_1 {}_F\mathfrak{I}\phi$. Let $K_1$ be an aspherical space with fundamental group $G_1$, containing $K_F$ in two ways, both representing $F \subset G_1$ and $\phi(F) \subset G_1$. Let $X = K_1 \cup (K_F \times [-1, +1])$, identifying $K_F \times (-1)$ with $K_F$ in $K_1$ and $K_F \times (+1)$ with $K_{\phi(F)}$. Define $A = K_F \times 0$.

It is an easy matter to see that $A$ is bicollared in $X$ in either case and that $X$ is aspherical, cf. [31], and $\pi_1(X) \approx G$, cf. [5] and [7].

There is some map $f : M'' \to X$ inducing an isomorphism of fundamental groups. By 2.B.2, $f$ is homotopic to $g$ in such a way that $g^{-1}(A)$ is a reduced bicollared 2-manifold in $M''$, and, for each component $T_i$ of $g^{-1}(A)$, $\pi_1(T_i) \to \pi_1(A)$ is injective. Thus each $\pi_1(T_i)$ is finite, and so $T_i$ is either a 2-cell, a 2-sphere, or a projective plane.

If $T_i$ is a 2-cell, then $\partial T_i \subset \partial M''$ must be contractible on $\partial M''$ and so must bound a 2-cell there. Let $T_i$ be a 2-cell for which $\partial T_i$ bounds an innermost 2-cell on $\partial M''$. By modifying $g$ in the neighborhood of the 2-cell in $\partial M''$, we can change $T_i$ to a 2-sphere.

If $T_i$ is a 2-sphere contractible in $M''$, then $T_i$ bounds a contractible submanifold $N \subset M''$, and a homotopy in the neighborhood of $N$ will remove $T_i$ from $g^{-1}(A)$.

If $T_i$ is a 2-sided projective plane, then it carries a nontrivial element of $\pi_2(M'')$ since the lifted 2-sphere in the orientable double covering of $M''$ cannot bound any contractible manifold—the covering translation would be a fixed-point free homeomorphism on this contractible manifold, contradicting the Lefschetz Fixed-Point Theorem.

Thus we have either the conclusion of the theorem in the end, or else we can find a map $h : M'' \to X$ homotopic to $f$ which has $h^{-1}(A) = \emptyset$; this is impossible since in such a case $h_*$ on $\pi_1$ would not be onto.

## REFERENCES


1. Poincaré, H., Cinquième complément à l'analysis situs. Rend. del Circ. Mat. di Palermo 18 (1904), 45–110.

2. Dehn, M., Über die Topologic des dreidimensionalen Raumes. Math. Ann. 69 (1910), 137–68.

3. Schreier, O., Die Untergruppen der freien Gruppen. Hamburg. Abh. 5 (1927), 161–83.

4. Kneser, H., Geschlossene Flächen in dreidimensionalen Mannigfaltigkeiten. JB. Deutsch. Math.-Verein. 38 (1929), 248–60.

5. Seifert, H., Konstruktion dreidimensionaler geschlossener Räume. Berlin Verb. Sachs. Akad. Leipzig, Math.-Phys. Kl. 83 (1931), 26–66.

6. Freudenthal, H., Über die Enden diskreter Räume und Gruppen. Commentarii. Math. Helv. 17 (1945), 1–38.

7. Kampen, E. R. van, On the connection between the fundamental groups of some related spaces. Amer. J. Math. 55 (1933), 261–67.

8. Grushko, I., Mat. Sbornik N.S. 8 (1940), 169–82.

9. Neumann, B. H., On the number of generators of a free product. J. London Math. Soc. 18 (1943), 12–20.

10. Hopf, H., Enden offener Räume und unendliche diskontinuierliche Gruppen. Comm. Math. Helv. 15 (1943), 27–32.

11. Waerden, B. L. van der, Free products of groups. Amer. J. Math. 70 (1948), 527–28.

12. Specker, E., Die erste Cohomologiegruppe von Überlagerungen und Homotopie-Eigenschaften dreidimensionaler Mannigfaltigkeiten. Comm. Math. Helv. 23 (1949), 303–33.

13. Moise, E. E., Affine structures in 3-manifolds. V. The triangulation theorem and Hauptvermutung. Ann. Math. 56 (1952), 96–114.

14. Papakyriakopoulos, C. D., On solid tori. Proc. London Math. Soc. 317 (1957), 281–99.

15. Papakyriakopoulos, C. D., On Dehn's Lemma and the asphericity of knots. Ann. Math. 66 (1957), 1–26.

16. Wagner, D. H., On free products of groups. Trans. Amer. Math. Soc. 84 (1957), 352–78.

17. Whitehead, J. H. C., On finite cocycles and the sphere theorem. Colloq., Math. 6 (1958), 271–82.

18. Shapiro, A., and J. H. C. Whitehead, A proof and extension of Dehn's Lemma. Bull. Amer. Math. Soc. 64 (1958), 174–78.

19. Bing, R. H., An alternative proof that 3-manifolds can be triangulated. Ann. Math. 69 (1959), 37–65.

20. Stallings, J. R., On the loop theorem. Ann. Math. 72 (1960), 12–19.

21. Epstein, D. B. A., Ends, Topology of 3-Manifolds, and Related Topics. Prentice-Hall (1962), 110–17.

22. Epstein, D. B. A., Projective planes in 3-manifolds. Proc. London Math. Soc. 11 (1961), 469–84.

23. Stallings, J. R., Some topological proofs and extensions of Grushko's Theorem. Thesis, Princeton Univ. (1959).

24. Stallings, J. R., A topological proof of Grushko's theorem on free products. Math. Zeitschr. 90 (1965), 1–8.

25. Stallings, J. R., A remark about the description of free products of groups. Proc. Cambridge Philos. Soc. 62 (1966), 129–34.

26. Waldhausen, F., On irreducible 3-manifolds which are sufficiently large. Ann. Math. 87 (1968), 56–88.

27. Stallings, J. R., On torsion-free groups with infinitely many ends. Ann. Math. 88 (1968), 312–34.

28. Swan, R. G., Groups of cohomological dimension one. J. Algebra 12 (1969), 588–610.

29. Eilenberg, S., and T. Ganea, On the Lusternik-Schnirelmann category of abstract groups. Ann. Math. 65 (1957), 517–18.

30. Serre, J.-P., Sur la dimension cohomologique des groupes profinis. Topology 3 (1965), 413–20.

31. Whitehead, J. H. C., On the asphericity of regions in the 3-sphere. Fund. Math. 32 (1939), 149–66.

32. Baer, R., Free sums of groups and their generalizations III. Amer. J. Math. 72 (1950), 647–70.

INDEX